THE U.S. NAVAL INSTITUTE ON

ARCTIC NAVAL
OPERATIONS

U.S. NAVAL INSTITUTE
WHEEL BOOKS

In the U.S. Navy, "Wheel Books" were once found in the uniform pockets of every junior and many senior petty officers. Each small notebook was unique to the Sailor carrying it, but all had in common a collection of data and wisdom that the individual deemed useful in the effective execution of his or her duties. Often used as a substitute for experience among neophytes and as a portable library of reference information for more experienced personnel, those weathered pages contained everything from the time of the next tide, to leadership hints from a respected chief petty officer, to the color coding of the phone-and-distance line used in underway replenishments.

In that same tradition, U.S. Naval Institute Wheel Books provide supplemental information, pragmatic advice, and cogent analysis on topics important to all naval professionals. Drawn from the U.S. Naval Institute's vast archives, the series combines articles from the Institute's flagship publication *Proceedings*, as well as selections from the oral history collection and from Naval Institute Press books, to create unique guides on a wide array of fundamental professional subjects.

THE U.S. NAVAL INSTITUTE ON

ARCTIC NAVAL OPERATIONS

EDITED BY TIMOTHY J. DEMY

NAVAL INSTITUTE PRESS
Annapolis, Maryland

Naval Institute Press
291 Wood Road
Annapolis, MD 21402

Library of Congress Cataloging-in-Publication Data
Names: Demy, Timothy J., editor. | United States Naval Institute, issuing body.
Title: The U.S. Naval Institute on Arctic naval operations / Timothy J. Demy.
Other titles: United States Naval Institute on Arctic naval operations |
    Proceedings of the United States Naval Institute.
Description: Annapolis : Naval Institute Press, 2019. | Series: U.S. Naval Institute
    wheel books | Includes index.
Identifiers: LCCN 2019022938 (print) | LCCN 2019022939 (ebook) |
    ISBN 9781682474792 (paperback) | ISBN 9781682474853 (epub) |
    ISBN 9781682474853 (pdf)
Subjects: LCSH: Arctic regions—Strategic aspects. | Arctic Ocean—Strategic
    aspects. | United States. Navy—Foreign service—Arctic Ocean. | Sea-power—
    United States. | Sea-power—Russia (Federation) | Sea-power—Canada. |
    National security—United States.
Classification: LCC UA880 .U534 2019 (print) | LCC UA880 (ebook) |
    DDC 359/.03091632—dc23
LC record available at https://lccn.loc.gov/2019022938
LC ebook record available at https://lccn.loc.gov/2019022939

♾ Print editions meet the requirements of ANSI/NISO z39.48–1992
(Permanence of Paper). Printed in the United States of America.

27 26 25 24 23 22 21 20 19   9 8 7 6 5 4 3 2 1
First printing

CONTENTS

EDITOR'S NOTE

Because this book is an anthology containing documents from different time periods, the selections included here are subject to varying styles and conventions. Other variables are introduced by the evolving nature of the Naval Institute's publication practices. For those reasons, certain editorial decisions were required in order to avoid introducing confusion or inconsistencies and to expedite the process of assembling these sometimes disparate pieces.

Gender

Many of the included selections were written when the armed forces were primarily a male domain and so adhere to purely masculine references. I have chosen to leave the original language intact in these documents for the sake of authenticity and to avoid the complications that can arise when trying to make anachronistic adjustments. So readers are asked to understand the context and celebrate the progress that we have made in these matters in more recent times.

Author "Biographies"

Another problem arises when considering biographical information of the various authors whose works make up this special collection. Some of the selections included in this anthology were originally accompanied by biographical information about their authors. Others were not. Those "biographies" that do exist vary a great deal in terms of length and depth, some amounting to a single sentence pertaining to the author's current duty station, others consisting of several paragraphs that cover the author's career. Because of these uneven variables, and because as a general rule we are more interested in what these authors have to say than who they are or were, I have chosen to even the playing field by forgoing accompanying "biographies."

Ranks

I have retained the ranks of the authors *at the time of their publication.* As noted above, some of the authors wrote early in their careers, and the sagacity of their earlier contributions says much about the individuals, about the significance of the Naval Institute's forum, and about the importance of writing to the naval services—something that is sometimes underappreciated.

Other Anomalies

Readers may detect some inconsistencies in editorial style, reflecting staff changes at the Naval Institute, evolving practices in publishing itself, and various other factors not always identifiable. Some of the selections will include citation support, others will not. Authors sometimes coined their own words and occasionally violated traditional style conventions. *Bottom line*: with the exception of the removal of some extraneous materials (such as section numbers from book excerpts) and the conversion to a consistent font and overall design, these articles and excerpts appear as they originally did when first published.

INTRODUCTION

The United States is an Arctic nation, and its responsibilities and interests in the region continue to grow. From exploration to exploitation, the Arctic has been a region of global interest for more than two centuries to American, Russian, and European powers and, increasingly, to Asia-Pacific nations. The U.S. Navy first entered the region in the 1800s.

Commercial interest and military interest in the Arctic have never been higher. Whether one seeks to navigate the Arctic for commercial shipping, research, tourism, or national security purposes, the need to safely and speedily navigate the waters of the Arctic Ocean is as pressing today as it was in 1845 during the ill-fated quest of Sir John Franklin to find the Northwest Passage.

The United States has been part of ongoing scientific research in the Arctic for decades. That research continues in the present and has gained momentum with concerns about climate change, rising sea levels, and melting polar ice. The U.S. Coast Guard has also long had a presence in Arctic waters, dating to the late 1800s and the service of the USRC *Bear* in Alaskan waters and the ongoing efforts of the International Ice Patrol.

By 1861 and the start of the American Civil War, American whalers had been fishing in Arctic waters for years. This was an economic point not lost on the Confederacy and Captain James I. Waddell of the raider CSS *Shenandoah*, who captured twenty-four whaling ships and sank twenty in waters near the Bearing Strait.

The U.S. Navy's long-standing interest in the Arctic has met with mixed results. The 1879–81 expedition of the USS *Jeanette* (formerly HMS *Pandora*) met with disaster, but in August 1958, the USS *Nautilus* (SSN 571) became the first submarine to complete a submerged transit of the North Pole.

During World War II, Winston Churchill had called the Allied Arctic convoys to Russia "the worst journey in the world," and more than one hundred Allied ships were sunk while attempting to reach Murmansk and other Arctic ports in the Soviet Union. Alaska and the waters around it also became a battlefield during the Aleutian Islands Campaign from June 1942 to August 1943, during which almost 1,500 Americans died and the U.S. Navy lost two ships (USS *S-27* and USS *Grunion* [SS 216]) and had two ships heavily damaged (USS *Salt Lake City* [CA 25] and USS *Abner Reed* [DD 26]). More recently the Navy maintained a presence on Adak Island from 1950 to 1997, and the station there was the deployment home of many P-3 Orion crews and others on isolated tours.

Today a new race for the Arctic is on, and Arctic and non-Arctic nations are entering it, seeking to gain faster sea routes to markets and renew their mineral stockpiles from the Arctic seabed. Drilling for Arctic oil is not the only economic and political issue. Many nations in addition to the United States and Canada claim national interests in the region. Russia has an impressive Arctic maritime construction program, and China has voiced significant interest and intentions in the region.

The challenges for any nation or military operating in the area are significant. Weather and sea conditions are formidable, as is the remoteness of the region and sparse infrastructure ashore.

In Aleksandr Solzhenitsyn's famous novel *One Day in the Life of Ivan Denisovich* (1962), the central figure, Ivan Denisovich Shukov, is a prisoner in a Soviet gulag. He is a former soldier whose crime was to escape from the Germans who took him prisoner in 1943 and return to his own lines. He is sentenced to ten years' hard labor in a Siberian work camp. In a harsh environment in every sense, where making it through one day is a daunting challenge and remarkable achievement, he muses, "How can you expect a man who's warm to understand a man who's cold?" It is a question well worth asking as the Navy continues its Arctic operations. It is indeed an environment of many challenges. The articles presented in this volume offer much to consider and should give readers reason to pause as the Navy extends its Arctic legacy and increases operations in the region.

1 "THE EMERGING ARCTIC FRONTIER"

ADM Robert J. Papp Jr., USCG

Well versed in Arctic operations, Admiral Robert J. Papp Jr. served as commandant of the U.S. Coast Guard from 2010 to 2014 and subsequently as U.S. Special Representative for the Arctic from 2014 to 2017. Papp argues that the importance of the Arctic region is growing enormously. Changes in the Arctic environment and advances in technology, as well as changes in military and commercial activity in the region, have made the Arctic vital to U.S. national interests, economy, and security. Once a region deemed largely inaccessible, the Arctic is now navigable and seeing increases in commercial shipping, military operations, tourism, and industrial endeavors such as those of the oil and gas industry. Rich in resources, the region has become an international geopolitical magnet that requires more national assets than presently available to the U.S. Coast Guard and other military services and government agencies.

"THE EMERGING ARCTIC FRONTIER"

By ADM Robert J. Papp Jr., USCG, U.S. Naval Institute *Proceedings* (February 2012): 16–21.

As a maritime nation, the United States relies on the sea for our prosperity, trade, transportation, and security. The United States is also an Arctic nation. The Arctic region—the Barents, Beaufort, and Chukchi seas and the Arctic Ocean—is the emerging maritime frontier, vital to our national interests, economy and security.[1]

The Arctic Ocean, in the northern region of the Arctic Circle, is changing from a solid expanse of inaccessible ice fields into an increasingly navigable sea, attracting increased human activity and unlocking access to vast economic potential and energy resources. In the 35 years since I first saw Kotzebue, Alaska, on the Chukchi Sea as a junior officer, the sea ice has receded from the coast so much that when I returned last year the coastal area was ice-free. The shipping, oil-and-gas, and tourism industries continue to expand with the promise of opportunity and fortune in previously inaccessible areas. Experts estimate that in another 25 years the Arctic Ocean could be ice-free during the summer months.[2]

This change from "hard" to "soft" water, growing economic interests and energy demands, and increasing use of the seas for maritime activities by commercial, native, and recreational users demands a persistent, capable U.S. Coast Guard presence in the Arctic region. Our mandate to protect people on the sea, protect people from threats delivered by sea, and protect the sea itself applies in the Arctic equally as in the Atlantic and Pacific oceans and Gulf of Mexico and Caribbean Sea.

The difference is that in the rest of the maritime domain, we have an established presence of shore-based forces, small boats, cutters, and aircraft supported by permanent infrastructure and significant operating experience. Although the Coast Guard has operated in southern Alaska, the Gulf of Alaska, and Bering Sea for much of our history, in the higher latitudes we have little infrastructure and limited operating experience, other than ice-breaking. Historically, such capabilities were not needed. Year-round ice, extreme weather, and the vast distances to logistical support, prevented all but icebreakers or ice-strengthened ships from operating there. As a result, commercial enterprise on any significant scale was nonexistent. But the Arctic is emerging as the new maritime frontier, and the Coast Guard is challenged in responding to the current and emerging demands.

Resource-Rich Realm

The economic promise of oil and gas production in the Arctic is increasingly attractive as supply of energy resources from traditional sources will struggle to meet demand without significant price increases. The Arctic today holds potentially 90 billion barrels of oil, 1.6 trillion cubic feet of natural gas, and 44 billion barrels of natural gas liquids, 84 percent of which is expected to be found in offshore areas. This is estimated to be 15 percent of the world's undiscovered oil reserves and 30 percent of natural gas reserves. Oil companies are bidding hundreds of millions of dollars to lease U.S. mineral rights in these waters and continue to invest in developing commercial infrastructure in preparation for exploration and production, and readiness to respond to potential oil spills or other emergencies.[3] In August, the Department of the Interior granted Royal Dutch Shell conditional approval to begin drilling exploratory wells in the Beaufort Sea north of Alaska starting next summer. ConocoPhillips may begin drilling in the Chukchi Sea in the next few years. Also, Russia has announced plans for two oil giants to begin drilling as early as 2015, and Canada has granted exploration permits for Arctic drilling.[4] The fisheries and seafood industry in the southern Arctic region (the Bering Sea and Gulf of Alaska) sustains thousands of jobs and annually produces approximately 1.8 million metric tons' worth of catch valued at more than $1.3 billion.[5] Although subsistence-hunting has occurred in the higher latitudes for centuries, as waters warm, fish and other commercial stocks may migrate north, luring the commercial fishing industry with them.

As the Arctic Ocean becomes increasingly navigable it will offer new routes for global maritime trade from Russia and Europe to Asia and the Americas, saving substantial transit time and fuel costs from traditional trade routes. In summer 2011, two Neste oil tankers transited the Northeast Passage from Murmansk to the Pacific Ocean and onward to South Korea, and Russian Prime Minister Vladimir V. Putin pledged to turn it into an important shipping route.[6]

Resolving an Old Liability on the Rule of Law

Because of these opportunities and the clamor of activities they bring, a legally certain and predictable set of rights and obligations addressing activity in the Arctic is paramount. The United States must be part of such a legal regime to protect and advance our security and economic interests.

In particular, for the past several years there has been a race by countries other than the United States to file internationally recognized claims on the maritime regions and seabeds of the Arctic. Alaska has more than 1,000 miles of coastline above the Arctic Circle on the Beaufort and Chukchi seas.[7] Our territorial waters extend 12 nautical miles from the coast, and the exclusive economic zone extends to 200 nautical miles from shore (just as along the rest of the U.S. coastline). That's more than 200,000 square miles of water over which the Coast Guard has jurisdiction.

Below the surface, the United States also may assert sovereign rights over natural resources on its continental shelf out to 200 nautical miles. However, with accession to the Law of the Sea Convention, the United States has the potential to exercise additional sovereign rights over resources on an extended outer continental shelf, which might reach as far as 600 nautical miles into the Arctic from the Alaskan coast. Last summer [summer 2011], the Coast Guard cutter USCGC *Healy* (WAGB-20) was under way in the Arctic Ocean, working with the Canadian icebreaker *Louis S. St-Laurent* to continue efforts to map the extent of the continental shelf.

The United States is not a party to the Law of the Sea Convention. While this country stands by, other nations are moving ahead in perfecting rights over resources on an extended continental shelf. Russia, Canada, Denmark (through Greenland), and Norway—also Arctic nations—have filed extended continental-shelf claims under the Law of the Sea Convention that would give them exclusive rights to oil and gas resources on that shelf. They are making their case publicly in the media, in construction of vessels to patrol these waters, and in infrastructure along their Arctic coastline. Even China, which has no land-mass connectivity with the Arctic Ocean, has raised interest by conducting research in the region and building icebreakers.[8] The United States should accede to the Law of the Sea Convention without delay to protect our national security interests: sovereignty, economy, and energy.

Arctic Responsibility

Wherever human activity thrives, government has a responsibility to uphold the rule of law and ensure the safety and security of the people. The Coast Guard is responsible for performing this mission on the nation's waters, as we have done in parts of Alaska over our 221-year history.

Coast Guard operations in the Arctic region are not new. Nearly 150 years ago, we were the federal presence in the "District of Alaska," administering justice, settling disputes, providing medical care, enforcing sovereignty, and rescuing people in distress. Our heritage is filled with passages of Coast Guardsmen who braved the sea and ice in sailing ships and early steam ships to rescue mariners, quash illegal poaching, and explore the great North. World War II ushered in the service's first icebreakers. In 1957, three Coast Guard cutters made headlines by becoming the first American vessels to circumnavigate the North American continent through the Northwest Passage. That mission was in support of an early Arctic imperative to establish the Distant Early Warning Line radar stations to detect ballistic-missile launches targeting the United States during the Cold War.

The Coast Guard presence in southern Alaska, the Bering Sea, and Gulf of Alaska continues to be persistent and capable, matching the major population and economic concentrations and focus of maritime activities. The 17th Coast Guard District is responsible for directing the service's operations in Alaska with:

- two sectors
- two air stations
- twelve permanently stationed cutters and normally one major cutter forward-deployed from another area

- three small-boat stations
- six marine safety units or detachments
- one regional-fisheries training center
- five other major mission-support commands.[9]

We ensure maritime safety, security, and stewardship in the region by conducting search and rescue, fisheries enforcement, inspection and certification of ships and marine facilities to ensure compliance with U.S. and international safety and security laws and regulations, and preventing and responding to oil spills and other water pollution.

The Coast Guard strengthens U.S. leadership in the Arctic region by relying on effective partnerships with other federal, state, local, and tribal governments and industry members. We are working with other federal partners within the Department of Homeland Security, the military services and combatant commanders within the Department of Defense, the National Oceanic and Atmospheric Administration, and the Bureau of Safety and Environmental Enforcement within the departments of Interior, State, and Justice to achieve unity of effort within the interagency team at the port and regional level. And we rely on cooperation from international partners, be they permanent close allies such as Canada or our maritime counterparts in Russia and China, with whom we are developing ties.

Although we have lived and served in southern Alaska for most of the Coast Guard's existence, our access to and operations in northern Alaska on the North Slope have been only temporary and occasional, with no permanent infrastructure or operating forces along the Beaufort or Chukchi seas. There are no deepwater ports there.

However, the acceleration of human activity in the northern Arctic region, the opening of the seas, and the inevitable increase in maritime activity mean increased risk: of maritime accidents, oil spills, illegal fishing and harvesting of other natural resources from U.S. waters, and threats to U.S. sovereignty. Those growing risks—inevitable with growth of human activity—demand the Coast Guard's attention and commitment to meet our responsibilities to the nation.

Preparing to Lead

Our first challenge is simply to better understand the Arctic operating environment and its risks, including knowing which Coast Guard capabilities and operations will be needed to meet our mission requirements. Operating in the Arctic region presents challenges to personnel, equipment, and tactics. What would be normal cutter, boat, or aircraft operations almost anywhere else become more risky and complex. The climate can be one of extremes many months of the year, with continuous sub-zero temperatures and more hurricane-force storms each year than in the Caribbean. It's hard on equipment: Industrial fluids freeze, metal becomes brittle, and electronic parts fail. It's also hard on people, who must acclimate to exaggerated daylight and darkness, harsh weather conditions, limited services, and isolation from family.

One of the most significant challenges is the lack of Coast Guard infrastructure in key locations along the northern Alaskan coastline that will be needed to sustain even basic shore-based operations. Today we rely on partner agencies and industry to support any sustained operations. Cutters, aircraft, boats, vehicles, and people require constant mission support and logistics. We are already exploring requirements to establish temporary forward-operating bases on the North Slope to support shore-based operations, enabling temporary crews and equipment to deploy to support a specific operation, and then return to home station when complete.

We have been improving our understanding by increasing operations. We conduct regular Arctic Domain Awareness flights by long-range maritime-patrol aircraft along the North Slope and over the Arctic Ocean, assessing aircraft endurance and performance and monitoring maritime activity. Since 2008, we have conducted Operation Arctic Crossroads, deploying personnel, boats, and aircraft to small villages on the Arctic coast such as Barrow, Kotzebue, and Nome. While there, we test boats for usability at these high latitudes and conduct flight operations. We also work closely with the Army and Air National Guard and the Public Health Service to provide medical, dental, and veterinary care to outlying villages. In return, we learn from their expertise about living and operating in this environment. These services invest in deepening our partnerships with and understanding of local peoples.

Next, we must prepare by ensuring that Coast Guard men and women have the policy, doctrine, and training to operate safely and effectively in the northern Arctic region. We have relearned fundamental lessons in recent years about the need to be prepared when taking on new operational challenges. We will train personnel beyond qualification to proficiency to live and work for extended periods in the extreme cold and other harsh conditions there. We will ensure cutters, aircraft, boats, deployable specialized forces, and mission-support personnel have the equipment, training, and support they require to succeed.

Finally, we are working closely with other key federal partners to lead the interagency effort in the Arctic. The Coast Guard has significant experience and success with speaking the interagency language, bridging the traditional divides between military and law enforcement at the federal level, and synchronizing efforts between federal, state, local, tribal, and private-sector stakeholders. Simultaneously a military service, a law-enforcement and regulatory agency, and an intelligence-community member that is part of the Department of Homeland Security, the Coast Guard is in a unique position to exercise leadership in this emerging maritime frontier.

Prevention and Response

Coast Guard missions rely on the twin pillars of prevention and response. We will take actions to prevent maritime safety, security, and pollution incidents in the Arctic. In our regulatory role, we are working with the Department of the Interior to review oil-spill response plans and preparedness by the oil-and-gas and maritime industries prior to exploration activities, especially on the outer continental shelf. We are taking the lessons from the 2010 *Deepwater Horizon* disaster to ensure that type of incident does not happen again, especially in the Arctic. We regulate U.S. mariners and inspect vessel- and facility-security plans. When a marine casualty does occur, we will investigate and take appropriate action to prevent it from happening again.

As a law-enforcement agency, we will provide security in the ports, coastal areas, and exclusive economic zone to enforce U.S. laws governing fisheries and pollution, while ensuring the security of lawfully permitted activities, including energy exploration, in the region. We will deploy cutters, boats, aircraft, and deployable specialized forces—maritime safety-and-security teams, strike teams, dive teams—when the mission demands.

As a military service, we will enforce U.S sovereignty where necessary, ensuring freedom of navigation and maritime homeland security. The *Healy*— our only operational icebreaker—and other ice-strengthened cutters will patrol where they can safely operate to provide persistent presence on the high seas and maritime approaches to the United States.

We are developing and will execute starting summer 2012 an Arctic Maritime Campaign with the objective of establishing a path forward for the Coast Guard to meet our responsibilities to the nation in the Arctic. This campaign will:

- define the required mission activities for the Coast Guard in the northern Arctic region
- determine capabilities (personnel, equipment, facilities) necessary to plan, execute, and support operations there
- identify available resources for the mission and resource gaps
- fully prepare our service and Coast Guard personnel to safely and effectively operate there.

Initially, the Arctic Maritime Campaign will be a Coast Guard plan for service operations in coordination with other partners—a basic first step for any mission. From there, we will work to improve interagency coordination as activities and operations increase.

My years at sea taught me many life lessons; chief among those is vigilance, the art of keeping a weather eye on emerging challenges so that the service can adequately prepare and take early and effective action to prevent and respond to trouble. As I scan the horizon, one area demanding our immediate attention is the Arctic. America is a maritime nation and an Arctic nation. We must recognize this reality and act accordingly. The Coast Guard is working to do its part. For more than 221 years, we have overseen the safety, security, and stewardship of our nation's waters. Our challenge today is to ensure we are prepared with a Coast Guard capable and ready to meet our responsibilities in the emerging maritime frontier of the Arctic.

Notes

1. U.O. Oode, Section 4111, "'Arctic' defined," http://codes.lp.findlaw.com/us
   -code/15/67/4111.

2. RADM David W. Titley, USN, and Courtney C. St. John, "Arctic Security Considerations and the U.S. Navy's Roadmap for the Arctic," *Naval War College Review*, vol. 63, no. 2 (Spring 2010), pp. 35–48.

3. U.S. Geological Survey Circum-Arctic Resource Appraisal, http://pubs.usgs
   .gov.usnwc.idm.oclc.org/fs/2008/3049/fs2008–3049.pdf. "Climate Change in the Arctic: Beating a Retreat," *The Economist*, 24 September 2011, p. 100. "U.S. to Offer Oil Leases in the Gulf," *The New York Times*, 19 August 2011.

4. "Shell Gets Tentative Approval to Drill in Arctic," *The New York Times*, 4 August 2011. Arctic Economic Development Summit, Chukchi Exploration Activities, ConocoPhillips Alaska, 5 August 2011, www.nwabor.org/AEDSdocs /22–35ConocoPhillips.pdf. "Russia Embraces Offshore Arctic Drilling," *The New York Times*, 15 February 2011. "Russia, Exxon Mobil strike deal for Arctic offshore oil drilling," *Anchorage Daily News*, 30 August 2011. "PEW Study urges Canada to suspend Arctic oil exploration," Terra. Wire, 9 September 2011, www.terradaily.com/afp/110909155430.fnr4r8w9.html.

5. "The Seafood Industry in Alaska's Economy," 2011 update of the Executive Summary, www.marineconservationalliance.org/wp-content/uploads/2011 /02/SIAE%5FFeb2011a.pdf.

6. "Breaking the Ice: Arctic Development and Maritime Transportation," Iceland Ministry for Foreign Affairs conference, 27–28 March 2007, www.mfa .is/media/Utgafa/Breaking%5FThe%5FIce%5FConference%5FReport.pdf. "Neste oil ships operate successfully along the Northeast Passage," Neste Oil Corporation press release, 30 September 2011, www.reuters.com/article /2011/09/30/idUS136183+30-Sep-2011+HUG20110930. "On Our Radar: Putin Covets Northeast Passage," *The New York Times* Green Blog, 23 September 2011, http://green.blogs.nytimes.com/2011/09/23/on-our-radar -putin-covets-northeast-passage/.

7. Department of Defense, Report to Congress on Arctic Operations and the Northwest Passage, May 2011, www.defense.gov/pubs/pdfs/Tab%5FA%5F Arctic%5FReport%5FPublic.pdf.

8. "Group: China preparing for Arctic melt commercial opportunities," *USA Today*, 1 March 2010. "China to launch 8 Antarctic, Arctic expeditions," ChinaDaily.com, 25 September 2011, www.chinadaily.com.cn/china/2011 –09/25/content%5F13788608.htm.

9. U.S. Coast Guard, "Protecting the Last Frontier," www.uscg.mil/d17.

2 "ARCTIC MELT: REOPENING A NAVAL FRONTIER"

RADM David Gove, USN

Environment changes in the Arctic in recent years have increased the opportunities, as well as the national security concerns, for Arctic nations (and others). National governments increasingly are regarding climate change as a component of strategic and operational planning. Competing political and economic claims in the Arctic region are complex and run the gamut from fishing rights to sovereignty rights. The latter are compounded in complexity because the United States has not ratified the United Nations Convention on the Law of the Sea, leaving it the only Arctic nation and industrialized nation not to have done so. Even if the United States ratifies the convention, U.S. naval interests will expand in the increasingly ice-free Arctic. These interests require increased military capabilities and operations. The lack of ice-hardened surface ships in the U.S. Navy and no new ice-breakers in the U.S. Coast Guard mean the nation is unable to ensure complete maritime domain awareness and superiority in the region.

"ARCTIC MELT: REOPENING A NAVAL FRONTIER"

By RADM David Gove, USN, U.S. Naval Institute *Proceedings* (February 2009): 16–21.

Increasingly, strategic planners and defense experts are viewing the issue of climate change in terms of national security. Reduced water resources and agriculture output in parts of the world could lead to political instability, armed conflict, and the spread of extremist and ultra-nationalist ideologies. While there is strong disagreement over whether the changing climate is a long-term trend or a shorter-term anomaly, it is incumbent on military planners to be prepared for any contingency.

Some of the most dramatic evidence of a changing climate can be found in the Arctic. Over the last several decades, significant changes in the region have been observed and measured, including a decrease in thick, hard multi-year ice. An ice-diminished Arctic may open the possibility of shorter routes for commercial shipping; greater access to gas, oil, and seabed mineral resources; and expanded commercial fishing opportunities. For these reasons, human activity and interest in the area is increasing, and there is a growing awareness of the strategic importance and nature of this unique region.

With roughly 1,000 miles of Alaskan coastline in the Arctic, the United States has a variety of national interests in the region, and the impact of physical changes there raises issues with respect to national sovereignty, security, and defense. One means for addressing such issues is the Arctic Council, an intergovernmental forum, which has undertaken collaborative projects in research and environmental protection. Member nations include the United States, Canada, Russia, Norway, Denmark (with claims through Greenland), Sweden, Finland, and Iceland.

Competing claims dealing with the Arctic are often political in nature and have important implications. For example, in the summer of 2008 Canada announced that it would increase its military presence in the region, begin construction of a deep-water port on Baffin Island, establish a cold weather training base at Resolute Bay, and build six new ice-hardened ships to patrol the Northwest Passage. During the same period, Russia conducted strategic bomber flights over the area for the first time since the end of the Cold War.

Within the U.S. government, the changing polar climate has resulted in increased focus on areas of national interest ranging from management of natural resources and research to national defense. Climate change in the predominantly maritime environment of the Arctic may have important implications for the Navy and Coast Guard.

Documented Change

Indicators of measured change in the region include increased air and sea temperatures, decreased extent of sea ice, degraded permafrost, reduced glacial coverage over Greenland, increased atmospheric water vapor, decreased snow cover, and larger discharges from rivers. All of these factors have noteworthy impacts on the region's ecosystems.

For the Navy and Coast Guard, decreased sea ice in particular is a problem. While sea ice has gradually decreased over the last 50 years, this trend has been accelerating over the last two decades—satellite measurements have shown an average three percent decrease per year. A 2007 report on the findings of the Second Workshop on Recent High-Latitude Climate Change quantified the issue: the 2007 summer minimum extent of sea ice was 40 percent less than the minima of the 1980s. According to the National Ice Center, a joint Navy–National Oceanic and Atmospheric Administration–Coast Guard office that monitors and measures sea ice for navigation safety, the summer of 2007 witnessed the smallest extent of Arctic sea ice coverage in the observational record. The 2008 summer season, recorded as the second smallest sea ice extent, reinforces the decline documented over the past 30 years.

With less ice coverage over the ocean, a cycle is generated: as more water is open to absorb solar radiation, becoming warmer, this in turn melts more sea ice. This loop may greatly boost the rate of change. All current climate models project a continuing decline in sea ice extent. According to a report by the U.S. Arctic Research Commission, a median of model predictions calls for an additional 30 percent decrease by 2050 with a 40 percent decrease in ice volume.

Much of what we currently know about the bathymetry and hydrography of the Arctic Ocean comes from measurements made by submarines during the Cold War. As the most experienced explorers of polar waters, submariners provided some of the first indications that the polar environment was changing, reporting a surprising degradation of the ice thickness in the 1990s. In a program known as Scientific Ice Exploration (SCICEX), the Navy provided *Sturgeon*-class nuclear-powered attack submarines for collaborative scientific cruises carrying civilian specialists to the Arctic basin. From 1993 to 2000, six SCICEX missions allowed scientists to gather data on the physical and biological properties of the northern waters, with particular emphasis placed on understanding the dynamics of sea ice cover, circulation patterns in the water, and the structure of the Arctic Ocean's bathymetry.

More recently the Navy's Arctic Submarine Laboratory sponsored ICEX 2007, a month-long ice camp located on the edge of the perennial ice. Designed to test submarine operations beneath the ice, the camp included researchers from various institutions working to better understand the dynamic change taking place in the Arctic ice fields.

Preserving Sea Lines of Communication

The most apparent change in the environment has been the decline of perennial (multi-year) ice, especially evident along the Northern Sea Route (NSR). This navigable waterway connects the Pacific and Atlantic Oceans via polar waters along the northern coast of Russia. Commercial carriers choosing the NSR over more conventional mid-latitude routes will trim their passages significantly, much as the aviation industry has adopted trans-polar great circle routes to reduce flight time. An article published by the American Scandinavian Foundation noted the distance between Yokohama and Hamburg is almost 5,000 miles shorter by using the NSR, so it is not surprising that the commercial shipping industry has expressed great interest in the potential of Arctic transits.

Preserving freedom of navigation in the region is an important tenet of U.S. policy. The NSR, however, is a contested waterway, with Russian claims of sovereignty competing against U.S. and European Union insistence that it is an international strait available to all nations, subject to mutually recognized terms. Another potential transoceanic shipping route may be the Northwest Passage, which extends from the Atlantic through Baffin Bay and the Canadian Archipelago and into the Pacific by way of the Bering Strait. Canada claims sovereignty over the waters of the Canadian Archipelago, although the United States and the European Union claim that the Northwest Passage also constitutes an international strait which allows right of innocent passage.

The 2008 summer ice minimum opened up both the NSR and the Northwest Passage (defined by the World Meteorological Organization as having less than one-tenth ice coverage) for the first time in recorded history.

Aside from access and right of passage, the Navy and Coast Guard, in particular, must also be concerned with strategic choke points such as the Bering Strait, Canada's Queen Elizabeth Islands in the Northwest Passage, and Russia's Severnaya Zemlya and New Siberian Islands in the Northern Sea Route. These narrow passages offer some protection from persistent ice blockage, but they are also vulnerable to control or blockade by adversaries that would significantly disrupt potential commercial shipping and oil transport.

While melting of sea ice facilitates maritime transportation, the melting of the permafrost on land may necessitate it. Warming on land is already causing degradation of land infrastructure by turning snow- and ice-packed roads into impenetrable morasses. Furthermore, as the permafrost melts, support structures of the Alaskan oil pipeline and the Mackenzie Valley gas pipeline sink deeper into the ground causing stress breaks in the pipes and badly degrading paved roads. With reduced transportation options on land, maritime alternatives for shipping oil and other resources will likely become increasingly important.

Sovereignty Issues

International maritime law assigns a nation complete sovereignty over its territorial waters, defined as waters extending 12 nautical miles from the low water mark of the associated coastline, known as the Territorial Sea Baseline (TSB). Beyond the territorial waters, coastal nations have resource management and exploitation control over their Exclusive Economic Zone (EEZ), which extends

200 nautical miles beyond the TSB. In an area where the continental shelf extends beyond the EEZ, known as the Outer Continental Shelf, nations may claim limited sovereignty over resources in the seabed. In the Arctic, areas outside the jurisdiction of bordering nations are considered part of the Arctic Commons and are administered by the International Seabed Authority.

The United States and Russia have tried to reach bilateral agreement on the Bering Sea based on the "median line" principle, which uses a line drawn equidistant from the TSB of each nation. The nations' representatives delineated this line in 1990, but while the accord is provisionally in force, the Russian parliament has yet to officially ratify it. At stake are 18,000 square miles of the Bering Sea that Russian hardliners are reluctant to cede.

Under Article 76 of the United Nations Convention on the Law of the Sea (UNCLOS), jurisdiction may be claimed by a nation based on undersea features that are considered an extension of the continental shelf if the structure is geologically similar to their continental landmass. In a highly publicized event in August of 2007, a Russian deep submersible planted the Russian Federation flag on the seafloor at the geographic North Pole, claiming sovereignty of the Pole. This symbolic claim was based on the Russians' contention that the Lomonosov Ridge, an underwater land feature that transects the Arctic Ocean and passes within 200 nautical miles of the North Pole, is an extension of the Siberian landmass. Canada counter-claims the ridge as an extension of its landmass, and Denmark has made a claim based on its province, Greenland.

Russia's claim to the Lomonosov Ridge, including the North Pole, is currently being reviewed by the 21-country Commission on the Limits of the Continental Shelf, a body established by a provision in UNCLOS to make science-based recommendations to member states regarding claims to the continental shelf. The commission is advisory in nature and operates independently of the UN.

Unfortunately, the United States is not represented in that forum because it has not ratified UNCLOS. In fact, it is the only Arctic nation—and the only industrialized nation—that has not ratified UNCLOS. Despite strong support from the Department of Defense, the State Department, and the President, Senate ratification remains elusive. The issues of Arctic territoriality will ultimately be decided under the framework of UNCLOS, but the United States will be absent from many of these discussions unless the treaty is ratified.

Access to Resources

As a moderating environment provides more opportunities for access into the northern waters, the potential for tapping anticipated oil and gas reserves has caught the attention of the energy industry. According to the U.S. Geological Survey (USGS) there are currently some 400 oil and gas fields in the region's onshore areas, comprising about ten percent of the Earth's petroleum resources. One of the first far northern areas to be developed for oil production was Alaska's North Slope. This area is rich in oil, and the Alaskan pipeline runs from Prudhoe Bay on the slope to Valdez on the Gulf of Alaska. If the North Slope is any indication, offshore oil and gas in the region could be significant. Core samples from the Lomonosov Ridge suggest that some 55 million years ago semitropical waters rich in organic matter covered the Arctic basin, perfect conditions for the formation of vast hydrocarbon beds.

Last July [July 2008], USGS completed the Circum-Arctic Resource Assessment of undiscovered conventional oil and gas resources in all areas north of the Arctic Circle. Their findings suggest a potential 90 billion barrels of oil, 1,669 trillion cubic feet of natural gas, and 44 billion barrels of natural gas liquids may remain to be found there, of which approximately 84 percent is expected to be discovered in offshore areas. While this is the most comprehensive study to date, it remains a qualified speculation until oil companies begin to dig assessment wells.

Valuable mineral resources may also be contained in the ocean floor, including potentially significant amounts of high-grade manganese, copper, nickel, and cobalt, as well as diamonds and gold. For the moment, the area's seabed resources are too difficult to harvest to make them commercially viable, however, technological advances may change that.

Of more immediate interest, retreating ice opens up the possibility of reaching untapped commercial fishing stocks, particularly in the Bering Sea. Warmer water temperatures may also result in some northward migrations of fish species. Commercial fishing could develop into an important economic concern in the not-too-distant future.

National Security Challenges

U.S. naval interests will face new challenges in an increasingly ice-free Arctic with a strategic objective to understand potential threats to the United States from the maritime domain. As throughout the global commons, the U.S. Navy must be aware of activities that could be harmful to national security interests in a region that will, no doubt, see fewer barriers to access by potential adversaries in the future. National and homeland security interests pertinent to the U.S. Navy in the region would include early warning/missile defense; maritime presence and security; and freedom of navigation and over-flight.

Portions of this unique operating environment may, in fact, become a more conventional Navy operating environment. Nonetheless, patrolling this frontier will not be easy, and maritime patrol aircraft and ships will still confront a dangerous and demanding environment. During the Cold War, the Arctic climate was cold enough to preclude significant moisture from entering the atmosphere, thus limiting the number of foul-weather occurrences. Warmer temperatures there will allow more moisture to be absorbed by the atmosphere, facilitating the development of local-weather events and greatly increasing the risk of aircraft in-flight icing and ice accumulation on ship superstructures, which may also degrade weapon systems.

Surface ships operating in the area must be ice-hardened, or reinforced to withstand potential encounters with floating ice. For operating in areas of first-year ice, the American Bureau of Shipping's "Rules for Building and Classing Steel Vessels" requires strengthening of the bow and stern. Further consideration must be given to propellers, rudders, fin stabilizers, and bow-mounted sonars that would be particularly vulnerable to ice. Atmospheric ice accumulation on the superstructure may also necessitate adjustments to ship buoyancy and stability design.

The U.S. Navy has no ice-hardened surface ships, and all of its icebreakers were transferred to the Coast Guard in 1966. Currently the Coast Guard has three icebreakers, although the USCGC *Polar Star* (WAGB-10) and USCGC *Polar Sea* (WAGB-11) are the only two built to handle heavy ice. They were designed to steam continuously at three knots through ice up to 16 feet thick.

At 30-years of age, both are near the end of their service life, with *Polar Star* currently in caretaker status. According to the Coast Guard's Ice Operations branch, it would cost an estimated $400 million to refit her for an additional 25 years of service.

The Coast Guard's other icebreaker, the *Healy* (WAGB-20), is a lighter ship designed to break 4.5 feet of ice continuously at three knots. While primarily used for scientific missions, she is also capable of logistics, search and rescue, ship escort, environmental protection, and law enforcement missions. Since 2003, the *Healy* has made three deployments to survey the uncharted seafloor around the Chukchi Cap. Working in conjunction with NOAA's Office of the Coast Survey and the University of New Hampshire's Joint Hydrographic Center, the breaker's multi-beam sonar and sub-bottom profiler were used to better define the extent of the U.S. continental shelf.

The U.S. Antarctic Program also operates an ice-strengthened research ship, R/V *Nathaniel R. Palmer*. These four constitute the entire fleet of U.S. ships equipped for polar operations. A 2006 study by the National Research Council recommended the replacement of the Polar-class ships.

To date, no action has been taken on that recommendation, and the council estimates that it will take ten years to get from approval to commissioning.

A New Front

While there is no absolute certainty that the Arctic will continue to warm, there are important consequences if it does. Human activity there is already rapidly increasing. The region is primarily a maritime domain and the U.S. Navy of the future must be prepared to protect sea lines of communication supporting maritime commerce and other national interests—including national security—there. In addition to thinking through how we adjust our shipbuilding emphasis to support such operations, the Navy should also be thinking strategically about building the necessary infrastructure to provide logistic support for Arctic patrols, search and rescue capabilities, and shore-based support activities.

To ensure complete maritime domain awareness in the region, and to provide our forces a competitive advantage, it will be necessary to have comprehensive knowledge of the physical environment. Data must be obtained by a suite of remote sensors (satellites, radars), autonomous sensors (data buoys, unmanned vehicles), and manned sensors (shipboard, coastal observing stations). Computer-based ocean and atmospheric models must be adjusted to the geophysical peculiarities of high latitudes. Communication lines for data exchange and reach-back processing at high-performance computing production centers must be robust and reliable. To ensure safety of navigation, we will also need to conduct more high-resolution bottom surveys and increase the scrutiny we place on sea ice conditions.

The Navy relies on its international and interagency partners for assistance to ensure success of maritime domain awareness and maritime security missions. To meet the demands of national security in the changing northern environment, strengthening mechanisms for cooperation among the regional nations and U.S. agencies must remain a high priority. Like everywhere else in the world, sound national security in the Arctic will require strategic access, military mobility, safe navigation, unimpeded maritime transportation, improved homeland security, and responsible, sustainable use of ocean and coastal resources. International and interagency agreements and partnerships are vital to incorporating these essential elements into a viable national security policy and will be critical for resolving future naval challenges of a changing Arctic.

Commercial carriers using the Northern Sea Route rather than conventional mid-latitude routes will save thousands of miles of travel, akin to the aviation industry's use of trans-polar great circle routes.

3 "GET SERIOUS ABOUT THE ARCTIC"

LCDR Brian Moore, USCG

Melting sea ice, a retreating polar ice pack, and an ice-free Northwest Passage have created commercial activities that previously were prohibited for shipping companies and the nations they serve and represent. Lieutenant Commander Brian Moore contends the United States needs a reinvigorated operational framework with commensurate capabilities. In this essay he argues that the "government's presence in the polar regions has been more symbolic than effective." New technologies and expanded capabilities have the potential for greatly enhancing operations by the U.S. Navy and U.S. Coast Guard and for strengthening sustainment and attainment of U.S. national interests.

"GET SERIOUS ABOUT THE ARCTIC"

By LCDR Brian Moore, USCG, U.S. Naval Institute *Proceedings* (August 2012): 40–43.

Commercial opportunities are expanding exponentially in the polar regions as the ice pack retreats, and the need for commensurate U.S. government engagement is also increasing. Shipping companies are eager to capitalize on the savings of time and fuel made available by an ice-free Northwest Passage and Northern Sea Route linking Europe and Asia. Oil and gas companies have already iden tified massive reservoirs suitable for development should the waters become reliably passable. A number of Arctic countries are already harvesting finfish and shellfish from polar waters, and other non-Arctic nations are demonstrating their intent by building ice-strengthened and ice-breaking-capable ships to facilitate their ventures. Ecotourism is burgeoning as luxury ships provide comfortable access to exotic and pristine wildlife venues.

This rapid uptick in human activity in the Arctic crosses numerous U.S. national interests, especially where the country has sovereign rights. Among the concerns are maritime safety, national security, economics, and natural resources (including fisheries, oil, and gas), national defense, and border control. National Security Presidential Directive 66 outlines the intended U.S. policy in the Arctic. However, to date [August 2012] the government's presence in the polar regions has been more symbolic than effective, conducted almost exclusively by ice-breaking vessels—most recently the U.S. Coast Guard cutters *Polar Sea* (WAGB-11), *Polar Star* (WAGB-10), and *Healy* (WAGB-20). The Department of Defense operates an aircraft- and missile-detection system, and submarines are thought to also operate to an extent in the Arctic. But to achieve the goals of NSPD-66, a broad scope of additional action is urgently required.

What We Have

As commercial maritime operations continue to ramp up in the near term, the Coast Guard's existing resources cannot keep up with the needed levels of shipping oversight, marine casualty and incident response, maritime domain awareness (MDA), and national defense. Of the service's three icebreakers—heavy breakers *Polar Sea* and *Polar Star* and medium breaker *Healy*—only the latter is functional. The other two, because of age and years of restricted budgets, are now inoperable and in need of significant overhaul. The *Polar Star* is scheduled to be ready for operations in late 2013, after significant reactivation work to allow her to potentially operate for another seven to ten years, barring further major mechanical breakdown. The *Polar Sea* would need mechanical work costing millions just to limp back under way for a few years, but even that is not budgeted, as the ship is slated for scrapping later this year.

Therefore, the *Healy* is the only U.S. government surface vessel capable of operating in polar and sea-ice conditions. Construction of new icebreakers takes many years and would cost upward of $1 billion each—thus being of no help in the near term, and an unlikely appropriation in the current budgetary climate.

So what operational paradigm can the government implement that can possibly meet all our national responsibilities while also clearing budgetary constraints? Fortunately, increased maritime use of the Arctic Ocean will necessarily be tied to summers, and the predominant environmental factor of Arctic summers is perpetual sunlight. This powerful characteristic presents a unique opportunity for mission fulfillment and personnel support. Solar-powered activities could be greatly facilitated and avoid difficult logistical challenges that petroleum-powered activities entail.

Use Solar-Powered Drones

Recent developments in winged and aerostat (tethered, unmanned airship) technology and solar power could easily be combined to yield a persistent, low-cost MDA presence using high-altitude orbiting aircraft with appropriate sensors. In the short term, a number of unmanned aircraft such as the *Zephyr* or the high-altitude airship, with a combination of radar, FLIR, and VHF/UHF radio receivers, could provide total Arctic Ocean activity awareness at a very small cost, and with no deleterious effect on polar ice masses that would be impacted by ships patrolling in the traditional sense, actively breaking ice.

Such aircraft could remain in flight all summer without refueling. They could reasonably be managed from existing U.S. government facilities near Anchorage, Fairbanks, and Dutch Harbor, Alaska; and Thule Air Base, Greenland. Additionally, some or all of these aircraft could be flown south for the winter to provide coverage in the antarctic when the Arctic shipping season closed. Other elements of the Homeland Security Department are already fielding such assets to provide surveillance of U.S. borders and aid in customs enforcement.

While MDA is the critical first step, just knowing what is going on in the Arctic Ocean will not be adequate. True sovereignty also requires the ability to act. Because of its unique authorities, the U.S. Coast Guard remains the logical governmental arm to carry out much of the nation's policy interests at the poles. But to do so means actually getting personnel on station anywhere in the Arctic, both in U.S. territorial waters and throughout the U.S. exclusive economic zone. To accomplish this, we have three alternatives: watercraft, aircraft, and submarines.

Fix the Cutters

Icebreaking Coast Guard cutters would seem the obvious choice, until we consider the vastness of the area. As in antipiracy efforts in the waters off Somalia, the number of ships required to effectively patrol the region would simply be too high for icebreakers alone to ever achieve adequate coverage. Additionally, the appropriations gauntlet for a high number of these expensive ships would be daunting. In the short term, though, a comprehensive service-life extension of both existing Coast Guard heavy icebreakers is the only way to rapidly begin to assert sovereignty. Such a move would also represent a credible beginning of U.S. government projection of presence in the Arctic.

Because the *Polar Sea* has been slated for decommissioning, a solid repair estimate is not available; a realistic estimate for returning her to service for the period necessary to build a new ship ranges between $50 million and $100 million. To put this cost into perspective, the funds necessary for proper refurbishment of the existing cutters are about the same as one year's worth of royalties from just a few of the larger Gulf of Mexico oil-production wells.[1] After production begins from Arctic wells, the government can expect a similar additional revenue stream.

This is an important point in a constrained budget climate: the capabilities proposed here can be paid for from new revenues rather than taking from existing commitments. After that, multiyear funding for new icebreaking cutters should be established to ensure that two or more (ideally), or at least one new polar-capable ship per decade is produced. This will allow for maintenance of the existing ice-capable ships and provide for the establishment of shipyard expertise to produce the unique vessels. At roughly $1 billion each, a minimum annual funding appropriation of $100–$200 million can sustain the specialized capability.

Once this program is established, should the geopolitical or physical environment experience a further "climate" change that hastens the need for additional polar ships, it will be much easier to ramp up production from an existing base than to create the capability from scratch. From a taxpayer point of view, the return on the investment of $200 million—about 3 percent of Gulf of Mexico royalties and rents—is a strong U.S. sovereign presence along with the potential to compete for the Arctic's economic resources. It also means the United States will be able to continue to expand on cutting-edge research on the Arctic ecosystem, as well as provide expertise and the enforcement of environmental-protection measures.

Bring in Helos and Subs

With the high situational-awareness level that drone reconnaissance platforms will bring, along with the watercraft capacity proposed here, the third leg of a clear U.S. Arctic presence should be high-speed response aircraft such as the V-22 Osprey or CH-47 Chinook helicopter, which could effectively mobilize from existing bases near Anchorage, Fairbanks, and Thule when urgent needs develop.

High-speed, amphibious, or hover-capable aircraft can easily drop emergency dewatering pumps, life rafts, and exposure suits to people in distress, or conduct hoist recovery evolutions where necessary. They can also airlift in personnel and supplies sufficient to conduct U.S. sovereignty-related functions.

With some additional region-specific policy decisions, as in other operational areas they can also be armed in the event that illegal activities or issues pertaining to national defense or sovereignty should arise. Either the Coast

Guard could procure the Osprey organically, or potentially the aircraft could be operated by the U.S. Marines with Coast Guard officers on board as mission authorities (as is done now with joint Coast Guard/Navy patrol boats). The establishment of such a routine interagency operation arrangement would also greatly enhance U.S. government capabilities for national defense of the region. Other amphibious aircraft and potentially wing-in-ground-effect vessels could be even better suited for operations over ice and tundra, but are not as operationally ready as are the Osprey and Chinook.

Submarines also meet the need for a U.S. presence and projection of authority, albeit in a different way. Like icebreakers they can provide large numbers of personnel and equipment on location, but with much greater alacrity than any surface craft or cutter, and without the concern for excessively contributing to ice break-up or critical-habitat destruction. The stealth characteristics of submarines also have a powerful deterrent effect, making potential wrongdoers wary.

Obviously the only provider of submarines is the U.S. Navy, and any arrangement for using them for a governmental law-enforcement presence in the Arctic would have to be through close cooperation between the Coast Guard and Navy. However, similar collaboration on mission fulfillment is well established in counterdrug operations, and U.S. naval vessels are presumably familiar with operating in polar waters. Therefore, if policy makers choose to implement such an arrangement, no practical impediment exists.

New Rules for a Changing World

In addition to personnel and platforms, the Coast Guard will need to promulgate new and stronger operational regulations for commercial activities in the Arctic and along polar shorelines. In an area cruelly unforgiving of errors, oversights, or equipment breakdown—almost any failure—can rapidly spawn catastrophes and are simply unacceptable. Implementing comprehensive safety management regimes, including the International Maritime Organization's International Safety Management Code and the pending Polar Code, will be an absolute minimum requirement from the standpoint of safety of life, property, and environmental protection.

Accession to these treaties by the United States needs to be emphasized, and pre-voyage compliance-verification inspections must be mandated. Further, without Senate accession to the U.N. Convention on the Law of the Sea treaty (UNCLOS), all U.S. oil, gas, and other mineral exploration and production opportunities are severely compromised. The United States should immediately join the more than 160 nations that have ratified UNCLOS. Doing so will enable us to legitimize our claims to resources in areas of the continental shelf that extend beyond the 200-mile exclusive economic zone. To quote former President George W. Bush, who, like President Barack Obama, supports U.S. ratification of the convention: "It will give the United States a seat at the table when the rights that are vital to our interests are debated and interpreted."[2]

None of the technology or interagency operational arrangements proposed here is new or unproven. Among the milestones that have already occurred are the 2011 transit of the fully laden Russian Suezmax oil tanker *Vladimir Tikhonov* through the North Sea Route, the 2010 grounding on a shoal of the large-capacity passenger ship *Clipper Adventurer* east of Kugluktuk due to inadequate use of available technology, and the 2011 Russian Arctic Circle installation of an oil-exploration platform.[3] Canada drilled numerous wells in the 1970s and 1980s but found too little to justify production at the returns then available; today the climate, both financially and meteorologically, is far more favorable.

Using a multi-tiered approach that combines unmanned reconnaissance drones for MDA and command and control, refurbishment of existing ice-breakers and commitment to new vessels, and staged high-speed-response aircraft along with the ability to call upon Navy submarines, the U.S. government can rapidly and very cost-effectively equip and empower the Coast Guard to provide clear and unequivocal MDA, the assertion of U.S. sovereign interests, and the capability for casualty and incident response. All of these are critical to ensure the safety of lives at sea and the protection of our national interests and homeland security, while protecting the marine environment. Congress must hasten to act. More agile international and industrial interests are rapidly leaving us behind.

Notes

1. Bureau of Ocean Energy Management, "Proposed Outer Continental Shelf Oil and Gas Leasing Program, 2012–2017," U.S. Department of the Interior, November 2011, www.boem.gov/uploadedFiles/Propose%5FdOCS%5FOil %5FGas%5FLease%5FProgram%5F2012–2017.pdf.

2. Alaska State Legislature, "Findings and Recommendations of the Alaska Northern Waters Task Force," January 2012, www.housemajority.org/coms /anw/pdfs/27/NWTF%5FFull%5FReport%5FColor.pdf.

3. "Tanker Vladimir Tikhonov Completes Successful Northern Sea Route Transit in a Week," *The Maritime Executive*, 1 September 2011, www.maritime -executive.com/article/tanker-vladimir-tikhonov-completes-successful -northern-sea-route-transit-in-a-week; "TSB Report on Clipper Adventurer Grounding Reveals Broken Equipment, Questionable Decisions," *Nunatsiaq Online*, 27 April 2012, www.nunatsiaqonline.ca/stories/article/65674tsb%5 Freport%5Fon%5Fthe%5Fclipper%5Fadventurers%5Fgrounding%5Fre veals%5Fbroken%5Fequipme/; "Gazprom Towing Oil Rig to Arctic Circle," UPI.com, 19 August 2011, www.upi.com/Business%5FNews/Energy -Resources/2011/08/19/Gazprom-towing-oil-rig-to-arctic-circle/UPI -44991313755411/.

4 "COLD FRONT ON A WARMING ARCTIC"

Barry S. Zellen

If U.S. national interests and territory need to be defended in a future conflict in the Arctic, how will that best be accomplished at the strategic level? Barry S. Zellen argues that the most likely threat will come from an emboldened and resurgent Russia and that logic and geography therefore favor U.S. European Command (EUCOM) as the combatant command best suited for U.S. defense of the Arctic. The Arctic provides an emerging Asia-Europe sea bridge, and the polar thaw presents new opportunities for trade between Asia and Europe. Zellen contends that Greenland and Iceland serve as the North Atlantic gateway to the Arctic and that, just as Allies used the region during World War II to assist the Soviet Union, the gateway could be used today, through the efforts and resources of EUCOM, to tame the aggressive instincts of the awakening Russian bear.

"COLD FRONT ON A WARMING ARCTIC"

By Barry S. Zellen, U.S. Naval Institute *Proceedings* (May 2011): 44–49.

The most probable emergent threat to northern security emanates from a bolder, resurgent, resource-enriched Russia, currently the most assertive of the Arctic states and intent on leveraging the full economic and strategic potential of its vast northern lands and seas. The Arctic Ocean's coastal nations are Russia, Norway, Denmark (Greenland), Canada, and the United States. Non-coastal states are Iceland, Finland, and Sweden. Most of these are European, and the non-European Arctic states are NATO members with close historical, cultural, and strategic links to Europe. Only Russia's far east, Alaska's southern coasts, and Canada's far western province of British Columbia abut Pacific waters. For all these reasons, a strong case can be made for U.S. European Command (EUCOM) being best suited for the defense of the Arctic.

Northern Command, which is responsible for the defense of North America, seems a likely choice to some, though the North American Arctic remains the most secure part of the Far North, thanks in large measure to the sparse population and extreme isolation of Canada's northern archipelago. Pacific Command (PACOM), encompassing Alaska's North Pacific waters, seems likely to others. They note that the industrialized states of Northeast Asia have a strong economic interest in emerging trade routes across the top of the world, and that China, America's next most likely peer competitor, sees the Arctic through a Pacific lens, as did Japan a generation earlier. But widespread usage of northern shipping lanes remains a long way off, even though the Northern Sea Route, the Arctic Bridge between Murmansk and Churchill, and the famed Northwest Passage are already being used on a tentative, seasonal basis.

Geography seems to favor the European Command, since Russia owns by far the largest sector of Arctic coast and the shallowest Arctic continental shelf. As long as the region's waters thaw, Russia will have greater access to more of its long-hidden offshore resource wealth than will any other state. History also appears to favor the Arctic being viewed as part of EUCOM's area of operations, as the longest recent conflict in those waters was not the relatively brief World War II battle for the Aleutians, but the six-year Battle of the Atlantic.

Even the U.S. Maritime Strategy at the Cold War's end viewed the Arctic's undersea domain as primarily a route to contain then–Soviet Russia's fleet in its home waters, before it could menace North America. For these reasons, the key to a secure Arctic, at least while it remains seasonally frozen and regionally isolated, remains tied to the fate of Europe and the ambitions of its largest state: Russia.

Greenland and Iceland: North Atlantic Gateway

During the Battle of the Atlantic, 1939–45, efforts to assert command of the seas resulted in an ongoing naval clash between Allied and Axis sea powers. Convoys resupplying the United Kingdom and Lend-Lease runs to Murmansk traveled northeast past Newfoundland, through waters south of Greenland and Iceland, on their way to free Europe. Thus, the high North Atlantic and Arctic waters have long been viewed in terms of the Atlantic alliance.

After Denmark fell to the Nazis, the Germans eyed Greenland as their first stage of a route to invade mainland North America via the Gulf of St. Lawrence, through to upper Canada along the Great Lakes, in much the way Britain did during the War of 1812. Greenland's vulnerability resulted in America extending defense protection on behalf of the Danish government in exile, and this continued through the entire Cold War era, as Soviet naval power grew.

Whoever holds Iceland and Greenland seems destined to command the North Atlantic. The role of the Greenland-Iceland-UK gap during the Cold War for both Soviet and NATO naval strategy was central, though untested by war. The U.S. Maritime Strategy of 1986 likewise viewed the Arctic and North Atlantic as important areas for forward operations to contain the projection of Soviet naval power; critics feared the Maritime Strategy would destabilize deterrence, but in the end it helped reassure Europe that Soviet power was far less potent than Moscow wanted the world to believe.

An Emerging Asia-Europe Sea Bridge

In terms of economic potential, North Sea oil, the fisheries of the North Atlantic, and the sea lanes vital for transatlantic trade all illustrate this area's strategic-economic importance as a bridge connecting Europe and North America. As the Arctic thaws, the fisheries, natural-resource extraction efforts, and sea lanes will edge farther north into Arctic seas, eventually facilitating the emergence of an Asia-Europe sea bridge.

But the fundamental strategic relationship will remain the same, another reason it makes sense to view the increasingly navigable and economically integrated Arctic as an extension of the North Atlantic. With a polar thaw, Northeast Asian trading states will find a shorter direct route to markets in Europe, making the security of ports in the North Atlantic, and the sea lanes they inter-connect, even more crucial.

The Koreans, Chinese, and Japanese are considering shorter, safe shipping lanes to Europe over the top, and the Koreans have taken the lead with regard to commissioning a new generation of ice-hardened tankers, though the Rus-sians still dominate when it comes to heavy icebreakers. The prospect of con-necting Northeast Asian markets to Europe through an Arctic maritime bridge is compelling, but winters will bring new ice in the Arctic basin, limiting the year-round viability of such sea routes.

It is unlikely that we will see the center of gravity tip entirely toward the Pacific, particularly given the enduring transatlantic relationships that have been forged across centuries of trade, wartime and peacetime alliances, and the much less united strategic environment in Northeast Asia.

When transpolar shipping does become more frequent, we may find reason for PACOM and EUCOM to consider joint operations in the Arctic. Even with Asian states eyeing Arctic routes, the North Atlantic still features in most of their plans. Iceland could become a primary trans-shipment hub for Asian cargo ships, positioning the high North Atlantic to remain of critical strategic importance. That may be one reason Moscow helped bail out Iceland with a €4 billion loan when its economy collapsed in October 2008, hoping to expand Russian influence in the high North Atlantic and counterbalance the Scan-dinavian states that share maritime borders with Russia and have historically contained its naval influence.

Beyond Iceland, if Greenland were to become estranged from the West and pursue an unfriendly secession from Denmark, Moscow could find yet another island-nation open to courtship. That would certainly favor Russia's strategic position, putting pressure on the West and its command of the high North Atlantic. But for the moment, Greenland's independence movement is a friendly one, with Denmark's blessing.

The United States and NATO allies should cultivate warmer relations with all the micro-states and territories of the high North Atlantic and Arctic. Alaska and Iceland have especially close political ties, so this could be a good foundation.

Inuit Interests

Greenland may well be the key, since no one at this stage can predict where the loyalties of an independent Greenland will lie. Embracing the Inuit and their seal-hunting traditions would also go far to reduce tensions between them and the Europeans, who oppose this hunting and the fur trade, despite their long history of fur empires. More concerted confidence-building measures could help to ensure that the interests of the Inuit, and of the modern states that jointly assert sovereignty over their homeland, remain aligned.

This might, in turn, help thaw relations between Canada and the European Union, solidifying transatlantic rapport and boosting regional security. During the 6 February 2010 meeting of G7 finance ministers in the Canadian Arctic, Nunavut leaders generously offered their visitors a taste of northern cuisine, including a staple of their subsistence diet: seal meat. The ministers' undiplomatic decision to disrespect Inuit hospitality in Nunavut's capital city, Iqaluit, by refusing to attend a feast held in their honor was certainly not Europe's best moment.

The opportunity to restore a climate of friendship and trust may still be with us, but a serious effort is needed to mend fences with the still-disappointed Inuit. This may be why Secretary of State Hillary Rodham Clinton rebuked her Canadian counterparts for their exclusion of indigenous northerners from an A5 conference on the future of the Arctic in March 2010, calling upon her peers to provide the Inuit with a seat at the table. Canada's northern natives may be few in number, but they control many local economic and political levers, and their interests are now fully backed by Ottawa, their partner in land claims, self-government, and northern development.

It would not take much diplomatic savvy for the Russian bear to seize the opportunity and share with the Inuit tasty slabs of whale and seal meat, hoping to drive a wedge between the people of the Arctic and Europe. Secretary Clinton's overture was thus a well-timed preemptive move to ensure the West doesn't lose the North on her watch.

Russian Side of the Arctic

In its sector of the Arctic, Russia focuses on its vast, resource-rich, uniquely shallow continental shelf—which it wants the world to recognize as Russian territorial waters. The United Nations Convention on the Law of the Sea (UNCLOS) will likely agree. The 2007 diplomatic stunt placing a Russian flag beneath the North Pole was less a grab for the polar seabed than it was an assertion that there is a *Russian side* of the Arctic. Moscow would probably welcome the selection of the North Pole as the boundary point, as it was in the Cold War.

UNCLOS and the International Seabed Authority may, after all the claims have been filed and adjudicated, find that Canadian territorial waters extend past the pole into what Moscow views as its side—or that Russian waters extend to what many in the West perceive to be *our* side. It all depends partly on what Canada, Russia, and the United States can prove to be their respective continental-shelf extensions.

With its extensive and increasingly accessible Arctic continental shelf chock-full of petroleum in exploitable quantities, Russia has much to gain from a thaw and is rehabilitating its all-but-abandoned Northern Sea Route to bring the treasure to market. The strategic importance of this wealth to the country's economic resurgence also provides ample motivation for Moscow to ensure an adequate defense of its northern domain. It can no longer count on nature for a great wall of ice. This could increase security tensions along the old East-West fault line.

In April 2010, Russia and its Cold War rival Norway resolved long-simmering disagreements over their offshore boundary line, easing the way to the joint development of bountiful offshore petroleum resources. But economic collaboration can, and throughout history has, yielded to nationalist rivalries and even war between trading partners. In the end, the old East-West rivalry could resurface. This possibility reinforces the notion that the Arctic as a region, and a potential theater of conflict, fits logically into EUCOM's area of operations and its continuing mission of securing Europe from external threat.

Funding Patriotism

Just as Canadians have a powerful emotional attachment to their northern frontier, Russians view the Arctic as an extension of their heartland. It has been and remains their key to their survival, militarily and economically. The intensity of this attachment, and the strategic importance of the heartland, which saved the nation from Napoleon Bonaparte's armies as it did Adolf Hitler's, combine to define a vital national interest for Moscow.

More than the other littoral Arctic states, Russia is inclined to make full use of its Arctic assets, even though the post-Soviet economic collapse led to a decade-long abandonment of many mega-projects in the vast and now-rusting Russian Arctic, and of the maritime infrastructure along the Northern Sea Route. But in recent years, with higher commodity prices changing the calculus, Moscow has reversed course. There is a growing commitment to Arctic resources along with an awareness that Russia's destiny is tied to the North.

Already Arctic naval, land, and air exercises have shown the world that Moscow is serious about its ambitions there. Along Russian borders, where regional military deployments could appear to be more menacing, such activities could lead to a reemergence of historic tensions with neighbors, especially after the 2008 assault on Georgia. There can be little doubt that Russia would aggressively defend its Arctic interests if Moscow felt they were threatened.

Still raw is Russia's loss of empire, first with the sale of Alaska to the United States, which many in Russia still feel was nothing short of wholesale theft. The history of that transaction remains clouded by distrust. Moscow transferred sovereignty over Alaska to the United States in 1867, and the commercial interests of the Russian American Company were sold to Hutchinson, Kohl & Co. of San Francisco, which was renamed the Alaska Commercial Company—after decades of sacrifice and investment by explorers who risked much to colonize the high North Pacific. Many Russians were perplexed by the abandonment of Alaska, and some nationalists still include Alaska on their maps, even though this is largely symbolic and not necessarily a reflection of military ambition.

With the Soviet collapse, Russia became even smaller and more vulnerable when it lost its Central European, Central Asian, and Baltic empire. The remaining Arctic lands and seas are thus highly valued as a sacred part of Mother Russia, a key to its future, and one of its last sources of pride and hope. Having agreed to purchase new French warships and with more heavy icebreakers than all its neighbors combined, Russia may well emerge a predominant regional power in the high North.

While Russia was at the table at the Arctic Ocean Conference held in Ilulissat, Greenland, on 27–29 May 2008 and pledged to support international law and the UNCLOS mechanism, one wonders what Moscow would do if the world community sided with Canada or Denmark in terms of continental-shelf extensions, at Russia's expense. The resolution of the border dispute with Norway is a welcome sign of a more collaborative Russia, but political winds can change.

On the other hand, much like what Soviet president Mikhail Gorbachev proposed in his prescient Murmansk speech on 1 October 1987, the Arctic could become a testing ground for a new relationship between Russia and the West, and perhaps even a path toward eventual NATO membership. But if competition trumps cooperation in the end, the Arctic may become one of the first regions where a newly assertive Russia confronts the West. This is one more reason for which EUCOM will be drawn into the increasingly salient and challenging mission of securing the Arctic.

Pacific Power and Economy

Japan made a dramatic but tenuous grab in its militarist past for the high North Pacific, gaining possession of the Kuriles, Sakhalin, and, during the opening shots of World War II, the outer Aleutians as well. But Tokyo's far-northern reign was brief, and currently its ambitions are primarily defensive. Japan is no longer a major strategic player in the high North Pacific, owing to the defensive mission of the Japanese Maritime Self-Defense Force. But with some 110 major warships, it remains an important partner, particularly with regard to countering China's increasing naval power.

China has increased its Arctic activities, while at the same time expanding its naval aspirations and capabilities from brown to blue water. But its primary far northern ambition is most likely to establish a secure, dramatically short-ened, direct trade route to Europe, and to benefit from the increasing trade in Arctic natural resources that were formerly inaccessible.

These economic interests favor a less-aggressive position than that of Japan during World War II. At that time Japan viewed the region's resources less collaboratively, considering the high North primarily for strategic defense of its home islands and as a tactical diversion for the U.S. fleet during the Battle of Midway. China's assertion of greater naval dominance of the South China Sea has precipitated a vigorous reaction from its neighbors in partnership with the U.S. Navy, suggesting that it's unlikely that China will be able to dominate the high North Pacific as Japan once did. China will compete aggressively for resources, but it will likely do so as a member of the world economy. The country may seek to explore the Arctic, and in so doing demonstrate that it has become a great power with global capabilities, but it is not likely to threaten the region.

In sum, as Northeast Asia's populous industrial countries monitor the thawing Arctic, they see it primarily as a gateway to European markets and a new source of natural resources for their expanding economies—less as a target for military expansion. With these states thinking in terms of trade, they are unlikely to pose a strategic threat to the region or its security. Consequently, the Russian bear stands alone as the primary Arctic power, with intentions and capabilities that could conflict with those of the West.

Transatlantic relations and the security of the West, as well its continuing integration with economies of the industrialized Far East, will increasingly depend on ensuring the security of the Arctic. This suggests that EUCOM could be the right command in the right place to take the lead on these defense issues.

Like the Arctic, EUCOM's area of responsibility is adjacent to Russia, and this ensures their fates will remain linked for years to come. Being near an awakening bear, and having experience in taming its aggressive instincts, will be an important key to a secure and peaceful North. While hoping the bear can be subdued and enticed to become a friend, we must be prepared for aggression. EUCOM, whose mission since the darkest days of the Cold War has been to defend the West, has the experience to do both.

5 "IN THE DARK AND OUT IN THE COLD"

LCDR Magda Hanna, USN

The opening of shipping lanes in the Arctic, the result of the depletion of Arctic ice, is creating new opportunities and challenges for which the United States is not prepared. As a result the possibility of threats to U.S. sovereignty and the freedom of navigation is real. Acknowledgment of the importance of the Arctic region in peacetime and wartime is not new, but the changing conditions are new and rapidly creating an environment of uncertainty for naval operations. Long-term planning, training, and acquisitions by the Navy for Arctic operations are increasing yearly, yet there need to be more. Without immediate attention, the Navy and the nation will find themselves left, as the title of this article says, "in the dark and out in the cold." That will be an operational and strategic failure and tragedy.

"IN THE DARK AND OUT IN THE COLD"

By LCDR Magda Hanna, USN, U.S. Naval Institute *Proceedings* (June 2006): 46–51.

Free-drifting icebergs, shifting boundaries of pack ice, 24-hour darkness, subzero temperatures, icing on ships' equipment and superstructures, and a lack of dependable logistical support can make Arctic operations extremely dangerous for surface ships. Nevertheless, naval strategists should start planning for Arctic operations to prepare for what may soon become a new challenge to U.S. sovereignty and freedom of navigation.

What is prompting this need? A team of more than 300 scientists recently confirmed unprecedented changes occurring north of the Arctic Circle. The Arctic Climate Impact Assessment released in November 2004 describes these changes, including a 3% per decade northerly retreat of the ice line or extent. Ice is also getting younger. At the rate of 7% per decade, persistent or multiyear ice is disappearing.[1] Four decades of U.S. submarine Arctic transits and under-ice surveys confirm that ice thinned by 40% in just the last 20 years.[2]

While these changes may not appear immediately relevant to U.S. national interests, significant reductions in ice coverage may soon begin to hit home. Changing global commercial shipping transit lanes and increased energy exploration will make the Arctic highly important in coming years. Given these dramatic changes, the Navy must review its strategic policy in the Arctic.

U.S. Arctic Strategic Policy 1940s–1990s

Interest in the Arctic as a theater of operations and critical supply line is not new. As part of the Lend-Lease Act of 1941, U.S. Arctic convoys began supplying the Russian ports of Archangel and Murmansk with critical aircraft and military replenishments. These convoys endured harsh Arctic weather conditions as well as frequent attacks by German submarines and battleships.[3]

In addition, the U.S. Coast Guard was charged with patrolling the east coast of Greenland during World War II. The importance of Greenland to the fight against Germany lay in its strategic geographic location, its supplies of cryolite for the aluminum industry, and weather stations that could be used for European weather forecasts.[4] The 1940s proved to be a critical time for learning the strategic advantage of Arctic surface operations.

After World War II, the Cold War brought the Arctic to the front lines as the only barrier between the Soviet Union and the United States. During this period, the Navy's role in the region steadily increased, and the Coast Guard maintained a strategic presence. Navy submarines learned the strategic advantage of ice cover, as both high and low sound frequencies suffer great propagation losses when acoustic energy bounces off the rough underside of ice.[5]

Surface ships played a critical role in Arctic strategic operations after World War II, with both the Navy and Coast Guard supporting icebreaker assets until 1965. The Coast Guard maintained seven deep-water icebreaking-capable vessels through the 1980s. These vessels were used for logistical support to high-latitude ports, science missions, and to support national defense interests. In the 1990s Coast Guard polar icebreaking assets decreased to two, with their mission remaining instrumental to scientific research and maintaining an Arctic-rescue and environmental-protection capability.

Historically, the Arctic has been a mainstay of U.S. maritime strategy, but since the end of the Cold War, this focus has weakened. A changed focus could be expected with the fundamental geopolitical shift from a bipolar Cold War strategic environment to the uncertainty of the Global War on Terrorism. However, with predictions of decreasing ice. this strategic and tactical environment could change dramatically.

Arctic Transits and Security

Diminishing ice over the Arctic poses new commercial shipping opportunities and hence, a U.S. strategic interest. The Northern Sea Route (NSR) and Northwest Passage are two potential trans-oceanic shipping routes that are predicted by some to become routinely navigable as sea ice recedes.[6] These transit routes are claimed by Russia and Canada, respectively, and commercial shipping companies are likely to begin exploring the economy of using these passages in the near future.

For the past decade, Russian ship captains have transited NSR waters on a regional basis, but few ships have attempted to use the full extent of these routes, which incidentally include the Bering Strait between Alaska and Siberia. Overall, transit length savings for commercial ships choosing northern routes over the mid-latitudes are speculated to be around 40%. For ships too large for the Panama Canal using the route around Africa, the distance is reduced by

6,770 miles or 15 to 20 days.[7] Added to this shortened distance would be the significant saving of fees required for transiting the Suez and Panama canals. Clearly, these numbers are piquing the interest of the world's top container-ship companies.

On the other hand, Dr. Franklyn Griffiths, professor emeritus of Political Science and George Ignatieff Chair of Peace and Conflict Studies at the University of Toronto, has argued that ship captains would encounter prohibitive insurance costs and unpredictable ice conditions in the Northwest Passage that would make it an unfeasible transit opportunity.[8] However, despite the current prohibitive risk level, diminishing ice will alter the Arctic environment and change these fundamental risk assessments.

Arctic Territorial Claims

While enormous economic gains could result from seasonally navigable Arctic sea routes, they could be a source of future economic and political tensions. Russian commercial producers expect the volume of oil flowing through the NSR to increase from one million to a hundred million tons per year by 2015.[9]

Currently, Europe and the United States do not recognize Russian and Canadian claims that the Northwest Passage and NSR sea lanes are internal waters.[10] In addition, the abundant northwestern Russian fisheries in the NSR region have proved to be a source of tension with Norway.[11] The current small disputes over Arctic transit claims could escalate if this region becomes more navigable and thus more economically important.

Disputed international claims do not end at transits and straits through the Arctic. Since 1999, the United Nations Convention on the Law of the Sea (UNCLOS) Article 76 has allowed coastal nations ten years from the date they ratify the treaty to make claims over the ocean floor apart from their 200-mile exclusive economic exclusion zone and beyond the limits of their continental shelves.

Denmark just launched a $25 million surveying project to prove that Greenland is geologically connected to the Lomonosov Ridge.[12] If proved correct, the Danes may be entitled to claim a large chunk of the Arctic Ocean all the way to the North Pole under Article 76. Using this same geological connection with Siberia and the UNCLOS provision, Russia also submitted a claim in the Arctic to the North Pole, which covers nearly 45% of the Arctic Basin.[13] Norway, Canada, and the United States have questioned Russian territorial claims but are also pushing forward their own bathymetric surveying efforts.

Arctic Energy Exploration and Strategic Concerns

The economic and transportation benefits the Russians have already realized are directly linked to discovered oil and gas reserves under the ice cap. According to Professor Alexander Granberg, advisor to Russian President Vladimir Putin, "Because of the Northern Sea Route, the Arctic is the leading economic region of Russia. The Arctic will develop much more quickly than all of the rest of Russia."[14]

Specifically, Russia currently uses the NSR regionally and will use it more extensively to bring oil and gas out of the Arctic. These plans are materializing through the current contract of the Finnish Ship builder, Aker Finyards, Inc., for construction of 20 ice-strengthened tankers to move oil and gas out of Murmansk.[15]

Russia is not the only country to gain from Arctic oil- and gas-rich continental shelves. Multinational oil companies continue to conduct exploratory drilling and seismic surveying in numerous Arctic locations in Alaska's North Slope and the Beaufort Sea. The U.S. Department of Interior's Minerals Management Service recently announced the sale of leases to 9.4 million acres offshore of Alaska's northern coast in the Beaufort Sea.[16]

Asian Energy Demand

The world's growing need for energy is driving the link between economic and transportation benefits of receding ice in the Arctic and U.S. security concerns. Specifically, demand for oil in China and India is expected to grow by an average annual rate of 4% per year until 2020. This demand, combined with declining production elsewhere, will make Asia's foreign oil dependence grow from 69% in 1997 to 87% in 2020.[17] This, combined with Middle East volatility, could increase oil prices and make Arctic oil reserves more economically attractive. China's need for the natural gas produced in Siberian gas fields may provide new energy cooperative agreements between Russia and China, or a source of instability, as China's energy insecurity grows.

China's energy demand could fundamentally alter shipping patterns as well as contribute to future Sino-Russian tensions. According to an intelligence report prepared for U.S. Secretary of Defense Donald Rumsfeld, "China is building strategic relationships along the sea lanes from the Middle East to the South China Sea in ways that suggest defensive and offensive positioning to protect China's energy interests, but also to serve broad security objectives."[18]

China is currently attempting to secure sea lanes that would ensure a smooth flow of future oil supplies to match surging demand. In the case of an ice-diminished Arctic, the Chinese may abandon efforts to use the U.S.-patrolled Straits of Malacca and instead pursue northern routes. This would prove to be not only a shorter transit for the Chinese, but the United States would be unable to follow and patrol the transits with its current capabilities.

Naval strategy in the United States has traditionally centered on defending sea lanes and strategic oil transits. The Straits of Hormuz and Malacca currently see the highest volume of oil flow daily. Increased Asian demand for oil and natural gas could fundamentally change these patterns of shipping. If key transits develop to the north, the United States must be prepared to patrol and defend these sea lanes equally.

U.S. Capabilities for Arctic Operations

Without a polar capability, the United States is seriously hindering its ability to ensure its national sovereignty and enforce laws and regulations along its borders.[19] An assessment of the U.S. Navy's preparedness to operate in Arctic or cold weather conditions reveals large gaps in knowledge. A workshop conducted in 2001, titled "Naval Operations in an Ice Free Arctic," brought to light many of these emerging concerns. The final report highlighted the deficiencies in U.S. Navy and Coast Guard equipment and training in cold weather.[20]

Specifically alarming is the lack of understanding of the operation of communications gear and sensors in the Arctic. Iced-over equipment and weapon systems would cripple battle-space management, navigation, and weapons deployment. Lack of navigational aids and accurate bathymetry also would paralyze any attempts to deploy naval assets in this environment. Currently, no U.S. source of satellite-based synthetic aperture radar (SAR) sensors can monitor Arctic sea ice under all-weather, day and night conditions. The tracking of ice features will require routine high-resolution monitoring only afforded by SAR. Unfortunately, the United States must rely on SAR imagery obtained from foreign sources, including Canada and Europe.

Despite these gaps, as well as the current possibility of emerging threats, the Navy recently eliminated its formal program for polar research at the Office of Naval Research, which is leading to a drain in expertise of Navy-supported researchers. This budget was as high as $25 million in the mid-1990s, but with formal elimination of the High Latitude Program, Navy-supported investments in Arctic research are now restricted to approximately $1 million annually, with a probable decline to zero soon.

The largest gap in capability nationwide is likely ice-strengthened hulls. To date, the United States operates one light icebreaker leased by the National Science Foundation (NSF), the *Nathaniel B. Palmer*, for operations in Antarctic waters, while the Coast Guard has three icebreakers capable of operating in the polar regions, the light icebreaker USCGC *Healy* (WAGB-20) and two Polar-class heavy icebreakers, *Polar Sea* (WAGB-10) and *Polar Star* (WAGB-11). However, the state of disrepair of the Polar-class vessels forced the United States to lease the Russian icebreaker *Krasin* to complete its Antarctic resupply mission last year. For the 2005–2006 resupply mission, the *Krasin* is expected to take the icebreaking lead, with the *Polar Star* participating only in a standby mode.

Reduced current capabilities in the U.S. icebreaking programs can also have long-term ramifications and be detrimental to U.S. security. For example, the Polar icebreakers' budget authority has shifted from the Coast Guard to the NSF in the president's fiscal year 2006 budget, and it is expected to remain with the foundation through at least fiscal year 2007. Logistical support of the McMurdo Antarctic Station will be the primary mission of the *Polars* over scientific research.

Maintaining homeland-defense capabilities or defending national sovereignty in high latitudes may be a distant third in interest—if at all. The potential danger in this arrangement lies in the fact that, while the logistical and science mission components of icebreaking can be contracted to commercial or foreign assets, sovereignty cannot be leased.[21] If made permanent and expanded to the *Healy* and future icebreakers, such changes could lead to the United States' lacking in military-controlled ice-operating surface vessels in the polar regions for the first lime in more than 100 years.

Immediate Concern, or Threat on the Horizon?

Some may argue that the demand for surface ships to operate in the Arctic is far enough in the future to not be cause for current concern. However, other countries' recent attention to the Arctic as well as other threats in cold-weather regions should not be ignored. Many of the gaps in capability for and knowledge of the Arctic apply equally to a North Korean threat. Maximum ice edge extents in the Yellow Sea reveal that ice will be a major cause of concern and could impede operational capabilities.[22]

Other countries have already begun earnest Arctic research. China has built and deployed an icebreaker and conducted a number of exploratory missions to the Arctic, including one unannounced visit to Tuktoyaktuk in Northern Canada in 1999. Canada, having already realized the vast importance of Arctic operations to its military, conducted Operation NARWHAL in 2003, an integrated Arctic military exercise to test its ability to operate effectively in the Arctic.

Even if a credible threat cannot currently be fully defined in cold regions, the United States may be forced to deal with law enforcement, homeland security, or environmental disaster problems in the Arctic as it becomes more navigable and widely used. Protection of the indigenous peoples of the north is another global concern that may arise in the near future, as their subsistence will be dramatically affected by development and energy exploration. Given the current global interest in the Arctic, as well as rising future concerns, the United States needs to take another look at its cold-weather knowledge and capabilities in the context of long-term security threats.

With the high probability of a continued diminishing ice cover trend in the Arctic, as indicated by the September 2005 satellite observations showing the lowest sea ice extent on record, the region can no longer be ignored as a potential theater for military operations. Territorial claims, increased maritime access, and vast natural resources are anticipated to create the need for future Arctic naval presence.

All of these changes are within the next Program Objective Memorandum cycle. Thus, fiscal planning for these changes needs to occur now. The Navy should begin long-term planning, training, and acquisitions toward understanding and operating in this complex environment. Without this, the Navy's lack of preparation could leave the United States in the dark and out in the cold.

Notes

1. Four-year scientific study of the Arctic region conducted by more than 300 international scientists. Compiled by Robert Correll in *Arctic Climate Impact Assessment* (ACIA). Summary of ACIA report can be found at http://amap .no/acia/.

2. D. Andrew Rothrock, *Was sea ice quite thin in the 1990s? Yes.* Presented at the SEARCH open science meeting 27 October 2003, Polar Science Center–Applied Physics Laboratory, University of Washington, Seattle, WA.

3. Randy Wilson, "The Siberian Connection: American Lend-Lease to the Soviet Union in WWII," *The Dispatch*, Winter 1998, Vol. 23, No. 4, available from http://rwebs.net/dispatch/output.asp?ArticleID=56.

4. Stetson Conn, Rose C. Engelman, and Byron Fairchild, *Guarding the United States and Its Outposts* (Washington, D.C.: U.S. Army, 2000). Available from http://www.army,mil/cmh-pg/books/wwii/Guard-US/ch17.htm.

5. *The Arctic Ocean and Climate Change: A scenario for the U.S. Navy*, U.S. Arctic Research Commission, 2000, copies available at http://www.arctic .gov/publications.html.

6. ACIA Report (see full citation in footnote 1).

7. Richard F. Pittenger and Robert B. Gagosian, "Global Warming Could Have a Chilling Effect on the Military." *Defense Horizons*, December 2003, available at http://www.stormingmedia.us/28/2832/A283224.html.

8. Franklyn Griffiths, "Pathetic Fallacy: That Canada's Arctic Sovereignty Is on Thinning Ice," *Canadian Foreign Policy Journal*, Volume 11, Number 3, September 2004.

9. Dr. Garrett Brass, Arctic Research Commission.

10. Arctic Marine Transport Workshop Draft Report held at Scott Polar Research Institute, Cambridge University, 28–30 September 2004.

11. Douglas R. Brubaker, *The Russian Arctic Straits* (Leiden, Netherlands: Martin Nijhoff Publishers, 2004), summary available at http://www.fni.no/publ/polar.htm.

12. Andrew C. Revkin, "Jockeying for Pole Position," 10 October 2004, *New York Times* Sunday late edition, available at Lexis Nexis.

13. The Brookings Institution, Policy Brief #137, David Sandalow, *Law of the Sea Convention: Should the U.S. Join*, available at http://www.brookings.edu/scholars/dsandalow.htm.

14. Arctic Marine Transport Workshop Draft Report held at Scott Polar Research Institute, Cambridge University, 28–30 September 2004.

15. Lawson Brigham, Arctic Research Commission, personal conversation (January 2005).

16. U.S. Department of Interior, Minerals Management Service Press Release, 24 February 2005. Available at www.mms.gov.

17. M. Ogutcu, "China's Energy Security: Geopolitical Implications for Asia and Beyond," *Oil, Gas and Energy Law*, 26 January 2003, available at http:/www.gasandoil.com/ogel/samples/freearticles/article%5f15.html.

18. Bill Ridley, "China and the Final War for Resources," Feb. 9, 2005. Available at http://www.energybulletin.net/4301.html.

19. Personal conversation with Capt. Dan McClellan, chief, Office of Strategic Analysis, U.S. Coast Guard, and a former National Security Fellow at Harvard University's John F. Kennedy School of Government (March 2005).

20. Naval Operations in an Ice-free Arctic Symposium Final Report (17–18 April 2001), report available at http://www.natice.noaa.gov.usnwc.idm.oclc.org/icefree/FinalArcticReport.pdf.

21. Capt. Dan McClellan (U.S. Coast Guard), personal communication (March 2005).

22. National/Naval Ice Center Archives (January 2004), contact www.natice.noaa.gov for further information.

6 "ARCTIC DOUBLESPEAK?"

Christian Le Miére

In the Arctic the convergence of national and regional interests, often with competing demands, has created a confusing and uncertain bureaucratic world. As Arctic activity increases, it is often difficult to discern whether the polar nations desire militarization of the region or security that discourages militarization. There is cooperation, but there is also increased political and military positioning wherein verbal and nonverbal signals are being sent but with unclear reception. There is military competition in the region, but nations also are restricted by decreasing budgets and scarce resources. The result is national desires for security, but strategic uncertainty.

"ARCTIC DOUBLESPEAK?"

By Christian Le Miére, U.S. Naval Institute *Proceedings* (July 2013): 32–37.

On 15 May [2013], ministers from the eight Arctic states gathered in Kiruna, northern Sweden, for their biennial meeting under the Arctic Council. The meeting was an excellent example of the difficult separation between securitization and militarization in the High North. While ministers used the meeting to sign the Marine Oil Pollution Preparedness and Response agreement, which highlighted the necessity for maritime safety and security in the region, there is also a growing military presence in the High North of assets that seem ill-suited to maritime-security operations. The Arctic Council specifically omits discussion of security and defense issues from its mandate, largely to avoid the potentially contentious issues of Arctic military deployments, but the lack of discussion furthers the idea that this is a competitive militarization of the region.

This raises the question of whether increasing military presence in the Arctic is driven by a desire to ensure safety and security within a wider international cooperative architecture, or by a desire to secure national sovereignty claims in disputed areas and insure against the activities of erstwhile enemies and current rivals. In short, is the growing military footprint in the Arctic a case of militarization or simply securitization?

Northern Exposure

The increased military deployments to the Arctic come amid rapid climatic change in the region that is creating newfound opportunities in hydrocarbon exploitation, mineral extraction, fishing, trade, and tourism. Seasonal sea ice in the Arctic is retreating at a rate far greater than previously expected. Minimum sea ice extent in 2012 was reached in September at approximately 1.32 million square miles, the lowest level since 1979 and almost 50 percent smaller than the 1979–2000 average.[1]

As a result, more areas of the Arctic are open to navigation than have been for perhaps thousands of years. This brings with it the possibility of new and more efficient trading routes between Asia and Europe. For centuries, Europeans sought the mythical Northwest Passage, finally conquered by the indefatigable Roald Amundsen in the early 20th century. However, it is the Northern Sea Route (NSR) along Russia's extensive coastline that presents the more immediate opportunity. In late 2012, the *Ob River* became the first liquefied natural gas tanker to pass through the NSR traveling from Norway to Japan.[2] Traffic remains limited along the NSR but is growing: In 2012, 46 vessels sailed the entire route, compared to just 34 in 2011 and four in 2010. In the very long term, there may even be a more direct route over the North Pole during summer.[3]

These routes are much shorter than traditional paths through the Suez or Panama Canals. The route from Shanghai to Hamburg, for example, is 3,231 miles shorter through the Arctic than the Suez Canal. This has potentially significant cost savings for shipping companies, with up to three or four weeks saved on any particular Asia-Europe journey.

It is not just maritime transport that is set to benefit from the opening of the Arctic. The exploitation of the Arctic Ocean's riches will also bring other economic benefits to the littoral countries and international fishing companies. A March 2012 U.S. Geological Survey report suggested the Arctic Ocean could hold 66 billion barrels of oil and 237,000 billion cubic feet of gas.[4] The proportion of fish catches to the global total is currently low (at about 5 percent), but as sea ice increasingly retreats, warmer seas attract migrating stocks, and fishermen are able to operate for longer periods farther north, this resource is likely to become more plentiful as well.

Pole Positioning

All this potential commercial gain and the largely unregulated environment in the High North have led to a media narrative emphasizing state-based competition, with newspapers eager to emphasize the seemingly zero-sum calculations of nations striving to secure resources.[5] The reality is somewhat more complex. While such a narrative has been supported by occasionally belligerent rhetoric and an increase in military deployments, messages sent by politicians and cooperative exercises are conflicting.

On one hand, the changes in the Arctic appear to be encouraging a rivalry for maritime rights. Most infamously, in August 2007 a Russian research expedition planted a titanium flag on the seabed at the North Pole. Although not an official government act, this propagandistic deed suggested a competitive mentality akin to imperial-era territorial grabs. A number of territorial disputes in the Arctic exist, some of them dormant, such as the Canadian-Danish Hans Island disagreement, and some largely benign, such as the U.S.-Canadian Beaufort Sea dispute. But the flag-planting episode was related to the most contentious of the disputes: a trilateral disagreement between Canada, Denmark, and Russia over their extended continental shelves, particularly the Lomonosov Ridge.

Accompanying such events and disputes has been a steady increase in Arctic deployments and operations. Russia, long the primary military force in the High North, has led the five Arctic littoral states in terms of military developments. The region was used regularly by the Soviet Union for submarine deployment under the ice in the Cold War. Russian submarine activity here was a concern not only for the United States during the Cold War, but also Scandinavian countries that on occasion saw Russian submarines approach major population centers. In 1981, for example, a Soviet Whiskey-class submarine ran aground in Gåsefjärden, Sweden, and the 1982 Håsfjärden incident involved an intense sub hunt in the Stockholm archipelago for what was believed to be a Soviet submarine.[6]

The Northern Fleet, based in Severomorsk and along the Kola peninsula and White Sea, has been the most substantial of Russia's five naval organizations and home to just under two thirds of the country's submarine fleet, including three fourths of the currently operational ballistic-missile submarine (SSBN) fleet.[7] While it has seen sizeable reductions in its capacity since the Cold War, there is now a concerted effort to rebuild or at least rejuvenate their presence in the north. The first new SSBN in more than 20 years, the *Yury Dolgoruky*, entered the Northern Fleet in January, with more to follow in the *Borei* class. The 200th Motor Rifle Brigade in Pechenga will form the base of a new Arctic unit that will be established within the Northern Fleet by 2015. Airfields that have been abandoned since the fall of the Soviet Union may be brought back into operation, while a squadron of MiG-31s will be deployed to Rogachevo on the island of Novaya Zamlya in Archangelsk Oblast by the end of this year.[8]

Russian activity in the Arctic has already increased. In August 2007, Moscow renewed long-range aviation patrols to the Atlantic, Pacific, and over the Arctic Oceans. Strategic bomber flights along the Norwegian coast increased from just 14 in 2006 to 97 in 2008; although this declined in subsequent years, it returned to over 55 flights in 2012.[9] In March 2013, two Tu-22M3 Backfire bombers and four Su-27 multirole aircraft flew within 20 miles of Sweden's borders; the failure of the Swedish air force to scramble in response to the night-time exercises led to searing media criticism.[10]

Russia is hardly alone in pursuing military development in the north. In Scandinavia, Norway has been the most active country modernizing its military, as its resource-driven economy and concomitant defense expenditure has not suffered the same restrictions as elsewhere in Europe. Norway's five *Fridtjof Nansen*–class frigates—the last being commissioned in 2011—have introduced a substantially more capable surface combatant to the navy, supported by six fast and stealthy *Skjold*-class patrol craft. Together, these two classes have greatly improved Norway's ability to defend its coastline. The importance of the High North to Oslo was also reflected in its decision to relocate the National Joint Headquarters from Jåttå in southern Norway to Bodo in the north in August 2009.

Denmark, too, has embarked on its largest procurement program to date with its three *Iver Huitfeldt*–class frigates. Copenhagen's focus on the High North was exemplified by the formation of a joint Arctic Command in October 2012 based in Nuuk, Greenland.

Neighborly Relations

The various platforms procured by these Arctic littoral states suggest a strategic focus on state-on-state warfare, and hence a militarization of the region, rather than an investment in the maritime-security capabilities that would benefit the growing commercial interests in the High North. This sentiment was shared by Russian President Vladimir Putin in February when he noted, "The danger of the militarization of the Arctic also persists."[11]

In particular, Russia's revitalization of its Arctic presence and Norway's concern over its larger neighbor most resemble a form of competitive military procurement. The purchase of new SSBNs, nuclear-powered attack submarines, and *Mistral*-class amphibious-assault vessels does not suggest a primary concern with maritime safety and security, but rather a desire to project power into and beyond the Arctic Ocean. In the same sense, the development of missile-laden surface combatants in Norway and Denmark do not suggest a maritime-security role for the forces being developed in the Arctic.

There are, however, indications that these procurements do not necessarily indicate a destabilizing competition in the region. The slow procurement pace by most Arctic states suggests that purchases are unlikely to develop into a naval arms race. Denmark is not expected to witness any nominal growth in its defense budget until 2015. Swedish defense spending has declined from 2.5 per cent of GDP in 1998 to 1.2 per cent in 2012, and Finland's defense expenditure to GDP has remained largely constant for the past decade.[12] The United States has equally shown little appetite to invest in any particular Arctic-focused facilities or equipment, despite the issue garnering political attention. Its ice-breaking capabilities remain limited to one heavy icebreaker, the USCGC *Polar Star* (WAGB-10) and one medium, the USCGC *Healy* (WAGB-20).

Further, the recapitalization of the Russian Northern Fleet may be a concern to neighboring states, but it could equally be argued that it is largely designed to update clearly decrepit equipment, rather than build a more muscular Arctic presence to coerce neighbors. Russia has witnessed rapid increases in its defense budget and is undertaking a substantial revitalization of its fleet. However, this should be viewed in context: Russia's navy has suffered from 20 years of stagnation since the end of the Cold War. Since 1994, only five new major surface vessels have entered service, and overall, much of Russia's navy, particularly its submarine fleet, is in dire need of modernization. Russia's newest naval shipbuilding program has deprioritized the more ambitious and less necessary ships, such as the aircraft carrier and cruiser program, in favor of platforms that will enable Moscow to monitor and govern its waters, such as frigates and corvettes.

There does, however, seem to be a contradiction in rhetoric, diplomacy, and arms procurement among Arctic states. Russia's perceived naval modernization sits uneasily with its otherwise relatively accommodating Arctic strategy. The potential riches to be gained from the opening of the region are encouraging Moscow to seek collaborative solutions to problems, for example, the 2010 Barents Sea agreement between Russia and Norway was the result of 40 years of negotiation over maritime delimitation of potentially hydrocarbon-rich waters. Speaking in 2010, Putin suggested the Arctic should be a "zone of peace and cooperation," and that "all the problems existing in the Arctic, including problems over the continental shelf, can be resolved through an atmosphere of partnership."[13]

Norway, too, exhibits such behavior, as Oslo has pursued its naval modernization program in conjunction with a somewhat schizophrenic Arctic strategy. While the country urges closer cooperation with Russia, it also encourages greater NATO presence there through the Cold Response invitational exercises it has hosted since 2006. Since 2010, Norway has held joint exercises with Russia, through the bilateral Pomor series, reflecting Oslo's desire to build a more collaborative military-military relationship with Moscow even as it purchases high-end platforms and weapons.

Canada is perhaps the exemplar of this confusion between the rhetoric surrounding seemingly competitive arms procurement and the reality of limited military capabilities in the High North. On one hand, Ottawa appears to be pursuing a competitive militarization in the region. Both Canada and Norway are expected to be major customers for the F-35 multirole aircraft program, with Oslo expected to buy 52 and Ottawa up to 65 aircraft, yet such aircraft do not appear to fulfill a significant maritime safety and security role. In an unusually bellicose statement, in 2007 Canadian Prime Minister Stephen Harper asserted that "Canada has a choice when it comes to defending our sovereignty in the Arctic: We either use it or lose it. And make no mistake this government intends to use it."[14]

At the same time, the Canadian military has attempted to build a more regular presence in the Arctic. A 500-strong army response unit for the Arctic is likely to be formed, and the military expects to construct both a seasonally available deepwater docking/refueling facility at Nanisivik and a training center in Resolute Bay, where the airfield may also be upgraded. The armed forces have also held three recurring exercises in the far north since 2007: Operations Nanook, Nunalivut, and Nunakput are designed to assert Canadian sovereignty as well as increase Arctic readiness.

Nevertheless, Ottawa's security force's presence is still relatively modest, comprising Canadian Rangers, a small army reserve unit under Joint Taskforce North (JTFN), signals-intelligence facilities at Canadian Forces Station Alert, and a chain of radars that provides an early-warning system in the north. Part of the expansion of the security force presence in the High North is to add 300 rangers within JTFN to bring the total to approximately 1,900. Perhaps the most significant Canadian military development has been its plan to procure between six and eight ice-strengthened Arctic offshore patrol ships, to be launched in 2015 at a current cost of $3.1 billion. These vessels won't boast offensive weapon systems; instead they will be humbly armed offshore patrol vessels intended to bolster Ottawa's ability to secure its territorial sovereignty.

Examining rhetoric and action highlights a key aspect of the question over whether recent military developments represent a militarization or a securitization of the region: the importance of intent rather than capability. If the Canadian deployments are intended to deliver greater security for the growing number of commercial Arctic users and deliberately designed to avoid confrontation with other states, then it matters little whether there is an increase in military presence. Indeed, it is arguable that some of the recent procurement and deployments are perceived not only as symbolic assertions of sovereignty but also as necessary deployments of security forces to prevent the emergence of a large, ungoverned space. The increased military attention in the High North may therefore at least in part be a securitization of the region rather than a militarization.

Northern Cooperation

Where does this leave the debate over the militarization or securitization of the Arctic? In short, the answer is somewhat confused.

To a large extent, the growth in military capabilities is relatively restrained and driven more by the knowledge that the retreat of sea ice in the Arctic will create vast areas of water that will require governance as increased traffic will demand security and safety. This need not be a competitive procurement process; Russia's desire to use the Arctic for commercial purposes means that Moscow is more likely to perceive collaboration as in its interests. Monitoring of traffic through different exclusive economic zones along the Northern Sea Route, for example, would necessitate coordination among constabulary agencies and information-sharing.

Recent diplomatic and military-military developments indicate that the Arctic states are aware of the need for greater interaction on military and security issues. The only two binding agreements the Arctic Council has reached have been designed to encourage cross-border cooperation on maritime safety and security issues: search-and-rescue and marine oil pollution. From the second chiefs of defense meeting in Denmark in June to the American-sponsored Arctic Security Forces roundtables, such military-military ties act as a form of confidence-building in a region with little security architecture.

In reality, the Arctic is an excellent opportunity to build closer ties among the littoral states, and thus between Russia, NATO states, and historical rivals. As Yevgeny Lukyanov, a deputy secretary of the Russian Security Council, noted in January, "Russia needs to cooperate with other Arctic states in strengthening and defending its Arctic borders and in monitoring transportation routes."[15]

At the same time, it is also undeniable that some of the platforms being purchased owe more to traditional occupations of state rivalry than the principles of maritime security. There is therefore some level of military competition within the region, but it is restricted by austere budgets and a narrative that currently favors cooperation. The question is whether this will continue to be the defining plot of the Arctic.

Notes

1. "Arctic sea ice extent settles at record seasonal minimum," National Snow and Ice Data Center, 19 September 2012, http://nsidc.org/arcticseaice -news/2012/09/arctic-sea-ice-extent-settles-at-record-seasonal-minimum/.

2. Clifford Krauss, "Gas tanker completes Arctic sea journey," *The New York Times*, 6 December 2012, http://green.blogs.nytimes.com/2012/12/06/gas -tanker-completes-arctic-sea-journey/.

3. Trude Pettersen, "46 vessels through Northern Sea Route," *The Barents Observer*, 23 November 2012.

4. "An Estimate of Undiscovered Conventional Oil and Gas Resources of the World, 2012," U.S. Geological Survey Fact Sheet 2012-3802.

5. For examples of such headlines see, Gerd Braune, "Cold War in the Arctic? Countries Seek Piece of Pie," *Der Spiegel*, 23 March 2009, and Steve Connor, "Arctic Ice Melt Will Bring Frosty Relations as Nations Navigate Across North Pole," *The Independent*, 4 March 2013.

6. See Carl Bildt, "Sweden and the Soviet submarines," *Survival*, July/August 1983, 165–169, for an overview of these incidents.

7. *The Military Balance 2013* (London: International Institute for Strategic Studies /Routledge, 2013), 231–234.

8. Märta Carlsson and Niklas Granholm, *Russia and the Arctic: Analysis and Discussion of Russian Strategies*, Swedish Defense Research Agency, March 2013, 26–28.

9. Katarzyna Zysk, "Military Aspects of Russia's Arctic Policy: Hard Power and Natural Resources," in James Kraska, ed., *Arctic Security in an Age of Climate Change* (New York: Cambridge University Press, 2011), 86–87.

10. "Ryskt flyg övade anfall mot Sverige," *Svenska Dagbladet*, 22 April 2013.

11. "Putin sees strategic balance threatened," *United Press International*, 27 February 2013.

12. Gerard O'Dwyer, "Sweden's Military Spending to Rise?" *Defense News*, 1 February 2013.

13. James Brooke, "Putin Stresses Cooperation in Arctic Resources Disputes," *Voice of America*, 22 September 2010.

14. "Canada to strengthen Arctic claim," *BBC News*, 10 August 2007.

15. "Russia calls for Tougher Arctic Security," *RIA Novosti*, 21 January 2013.

7 "OCEAN GOVERNANCE IN THE HIGH NORTH"

Ambassador David Balton and RADM Cari Thomas, USCG

Ninety percent of global commerce transits by ocean, and shipping in the Arctic is increasing. With this increase come greater risks and the need to develop more international collaboration to achieve common goals, including preservation of the Arctic marine environment and management of its resources. Such cooperation is a necessity for the eight nations, including the United States, that have a border and geographic stake in the Arctic. Ambassador David Balton and Rear Admiral Cari Thomas highlight several governmental and nongovernmental entities and groups that are active in Arctic policy and in shaping the international legal environment in which the United States and others will operate. In addition to discussing security interests, safety, environmental protection, and fisheries management, they point to ongoing concerns that, if not addressed adequately, place the nation at risk on several levels.

"OCEAN GOVERNANCE IN THE HIGH NORTH"

By Ambassador David Balton and RADM Cari Thomas, USCG,
U.S. Naval Institute *Proceedings* (July 2013): 19–23.

Changing conditions in the Arctic Ocean, including diminished sea ice, warming sea surface temperatures, and increasing maritime traffic, may require affected governments to take new steps to enhance their collaboration in the region. Those same changing conditions may also require governments to devote greater assets to preserve maritime security and assist in energy, environmental, and economic security. Through the Arctic Council, some work has been done in discrete areas (search and rescue, as well as oil-pollution preparedness and response), but there is more to do. Here we propose possible additional areas for international collaboration toward a common goal: preserving and protecting the Arctic marine environment and its resources while managing the balance of increased development, economic benefit, and related security interests.

The prosperity of our nation and many others depends on our oceans. Healthy oceans and sound maritime governance make possible the dependable arrival of containers shipped by sea, the supply of fuels shipped in tanks, the enjoyment of passengers traveling to experience the wonders of nature, and the livelihoods of fishermen. According to the U.S. Navy's *A Cooperative Strategy for 21st Century Seapower*, 90 percent of the world's commerce travels by sea, and a vast majority of the world's population lives within a few hundred miles of the oceans. The global nature of trade by sea is forcing the international shipping communities to examine new sea routes to reduce the all-important commodity: time. As a result, traditional international routes are under review, and possibilities across the Arctic Ocean are under consideration. The result? The need to examine regimes, governance structures, the expanse of tools of national security, and how the sea lines of communication can be used are all critical to retain a leadership role for the United States.

Emerging Needs for a New Melting Frontier

Compared to other maritime areas, the Arctic Ocean remains largely undeveloped and poorly understood. At 5.4 square million miles, it is the smallest of the oceans, but in 2012 ice coverage fell to its lowest level—1.32 million square miles.[1] Although the nations of the Arctic have demonstrated a high degree of cooperation in recent years, much more work lies ahead. All agree that the trend in diminishing sea ice is likely to continue. The strategic questions then become: What considerations need to be addressed, and which of these are in our vital national interests?

Undiscovered oil and gas are but two of the types of resources that may be available; exact estimates are thought to include more than 13 percent of the world's oil and 30 percent of its gas. In Alaska, energy-sector employment comprises more than 35 percent of the state's total available jobs. Increases in revenues through tourism also may be on the rise, as more than 1.2 million visitors are expected to use the region in 2013, an upswing of more than 300 percent.

Because of the increase in open Arctic Ocean, access to shipping routes, energy and mineral reserves, and fisheries resources becomes possible. Just as communities need to plan for various land uses, nations must plan for the new and expanding ocean uses in the Arctic. Without this, the safety of the people that may transit the waters, the well-being of the coastal Arctic communities, the sustainability of the resources that the area provides, and the smart stewardship of the environment all can be compromised.

Since the early days of sailing ships, the world's oceans have been used for a number of purposes. Whether for exploration, transportation and trade, for sovereignty goals, or to sustain the citizens using the resources from the sea, ships have been the most efficient ways of accomplishing these ends. But with any adventure at sea comes risk, as sailors know all too well. Some of the world's worst maritime disasters involved ships engaged in all of those purposes.

Throughout history, nations have used ships as instruments of power, to transport troops for self-defense, or in the conquest of another nation's territory. In 415 B.C., Athens expended its remaining reserves in an attempt to beat the Spartans in the Peloponnesian War. In the Sicilian Expedition, the Athenian fleet included 134 ships and 130 supply boats, and more than 27,000 troops and crews. In the end, ships, soldiers, and sailors were lost.[2] In 1945, several more well-known maritime tragedies occurred: the loss of the *Wilhelm*

Gustloff, in which upward of 10,000 German soldiers and civilians died, the loss of the *Yamato* and 2,475 Japanese crewmen, and the loss of the USS *Indianapolis* (CA-35) and 883 sailors.

Similarly, ships engaged in trade and transportation have met disaster. Probably the most well known was the sinking of the *Titanic* in 1912, in which 1,517 passengers and crew died. This loss led to the adoption of the first International Safety of Life at Sea convention—SOLAS, outlining international rules for the safety of ships and their passengers and crew. As transoceanic shipping grew in the early 20th century, so did the need for better governance to set global standards and requirements. The International Maritime Organization (IMO) was established and functioning by 1959. Topics of concern included carrying dangerous cargo, standardized measurements for tonnage, pollution, and search and rescue.

Hands Across the (Icy) Water

As human activity in the Arctic began to grow, the nations of the region saw the need to strengthen international coordination on Arctic issues. The Arctic Council, created in 1996, is the only diplomatic forum focused solely on the entire region. Its membership is limited to the eight nations with land territory above the Arctic Circle: Canada, Denmark (via Greenland), Finland, Iceland, Norway, Russia, Sweden, and the United States. Twelve other nations are currently accredited observers to the Arctic Council, with others seeking that status as well. A number of non-governmental observers—environmental groups, academia, and private-sector organizations—are either already participating as observers or wish to do so.

Six groups representing most of the indigenous peoples of the Arctic also have the status of "permanent participants" in the council. Although only governments have formal decision-making authority, the permanent participants in practice have very significant influence within the forum.

Six standing working groups of the council undertake projects on technical and scientific issues. Most projects are relatively low-key and non-controversial, while a small number are more politically significant and require negotiated outcomes. The most notable project thus far was the 2004 *Arctic Climate Impact Assessment* which took the first-ever comprehensive look at the impacts, both environmental and social, of climate change in the Arctic. Other major council products include the 2009 *Arctic Off-Shore Oil and Gas Guidelines* and the 2008 *Assessment of Oil and Gas Activities in the Arctic*. The council also created temporary bodies to carry out specific initiatives, including task forces to develop specific international agreements for the region.

Overseeing the work of the subsidiary bodies and managing the day-to-day operation of the council are eight Senior Arctic Officials (SAOs), one from each member nation. The SAOs are senior foreign-policy makers in their governments, and they advise their foreign ministers on Arctic Council matters.

The council meets at the ministerial level every two years to approve the work program for the biennium and to adopt any significant finished projects. Each of the eight Arctic nations holds the chairmanship of the council for two-year periods on a rotating basis. Sweden held the chairmanship until May 2013, to be followed by Canada (2013–15) and the United States (2015–17).

Protecting the Environment, Rescuing the Distressed

In 2011 the council completed its progress report on the 2009 Arctic Marine Shipping Assessment (AMSA), the most comprehensive analysis ever undertaken of trends relating to shipping into, out of, and through the region. The United States, Canada, and Finland led the AMSA project, based on a mandate in the Arctic Council 2004 Reykjavik Declaration.

The AMSA included recommendations in three areas: enhancing marine safety, protecting people and the environment, and building the marine infrastructure. These non-binding recommendations have helped set the agenda of the council, Arctic and non-Arctic states, international organizations, and shipping interests.

The Arctic nations have already implemented two of the key recommendations from AMSA, namely, the development of a search-and-rescue agreement and another on marine oil-pollution preparedness and response. A third recommendation—safety and environmental standards for shipping in both polar regions—is also moving forward through the IMO.

It was common practice among the earliest mariners to assist one another in distress. Eventually, customary international law required mariners to assist each other on the sea because of the time and space required to perform rescues. As the IMO has matured, developing regulations for preventing collisions at sea and assigning global responsibilities for the conduct of search-and-rescue coordination have been developed.

For all the changing conditions of the Arctic Ocean, one thing has not changed: the basic rules of international law relating to oceans. These laws apply to the Arctic in the same way that they apply to all the oceans. The international legal oceanic framework remains the 1982 United Nations Convention on the Law of the Sea (UNCLOS). The United States has not yet become party to it, despite the fact that we recognize its basic provisions as reflecting customary international law and follow them as a matter of long-standing policy.

Our status as a non-party to the UNCLOS, however, puts the United States at a disadvantage in a number of fundamental respects, most of which lie beyond the scope of this discussion. But our efforts to address the changing Arctic region bring at least two of those disadvantages into sharp focus.

First, we are the only Arctic nation that is not party to the UNCLOS. As our neighbors debate new ways to collaborate on Arctic Ocean issues, they necessarily will rely on the UNCLOS as the touchstone for their efforts. The United States will continue to take part in these initiatives, but our non-party status deprives us of the full range of influence we would otherwise enjoy in these discussions.

Second, the four other nations that border the central Arctic Ocean—Canada, Denmark/Greenland, Norway, and Russia—are advancing their claims to the continental shelf in the Arctic beyond 200 nautical miles from their coastal baselines. The UNCLOS not only establishes the criteria for claiming such areas of continental shelf, it also sets up a process to secure legal certainty and international recognition of the outer limits of those shelves. The United States also believes that it will be able to claim a significant portion of the Arctic Ocean seafloor as part of our continental shelf. But as a non-party to the UNCLOS, we place ourselves at a serious disadvantage in obtaining that legal certainty and international recognition.

Code of the North

As noted earlier, the IMO is developing a mandatory code for ships operating in Polar regions—the so-called "Polar Shipping Code." It will replace existing non-mandatory guidelines for ships operating in ice-covered waters that the IMO first approved for the Arctic in 2002 and revised in 2009 to include Antarctic waters. We anticipate that work on the code will continue through 2014.

We envision it to include both mandatory regulations and non-mandatory elements. The mandatory provisions will take the form of amendments to existing IMO instruments such as the Convention for the Safety of Lives at Sea (SOLAS), the International Convention for the Prevention of Pollution from Ships, and other relevant instruments. The new code will address unique hazards for ships operating in polar waters, taking into account factors such as the extreme operating conditions and remoteness of polar regions. Requirements such as ship-design and engineering standards (including hull-strength requirements), equipment, operations, manning and training (including the need for a qualified ice navigator in appropriate cases), and similar items will be included. These are often referred to collectively as the "safety chapters" of the Polar Code. It will also contain an "environmental chapter" addressing mitigation of potentially adverse environmental impacts of shipping operations.

The United States has joined many other IMO delegations in supporting the view that the safety chapters should be implemented through amendments to SOLAS or other appropriate instruments as soon as those chapters are completed and that implementation of them should not be unnecessarily delayed if the environmental chapter is still in development. However, we also believe that development of this chapter is too important to be put on hold while the safety chapters are finalized. We need to bring all aspects of the code on line promptly to deal with the significant increase in polar shipping that has already begun.

For several years, the United States has expressed concern about the lack of an international mechanism to manage potential fishing in the high-seas portion of the central Arctic Ocean. We have noted that vessels from any nation could begin fishing in this area in the near future. In 2008 Congress passed a joint resolution calling on the United States to initiate discussions with other relevant governments to address the situation. President George W. Bush signed the joint resolution into law as P.L. 110–243. In particular, the law envisions one or more international fisheries agreements for the management of Arctic fisheries as an ultimate objective. Until such an agreement or agreements are in effect, the law provides that "the United States should support international efforts to halt the expansion of high-seas fishing activities in the high seas of the Arctic Ocean."

Fisheries and the Future

The United States has also taken the unprecedented step of closing its own exclusive economic zone (EEZ) north of Alaska to any commercial fishing until domestic fisheries managers have sufficient information on the ecosystem of that area to allow fisheries to proceed on a well-regulated basis.

We also propose developing a new international agreement to cover the high-seas area of the Arctic Ocean beyond our EEZ. Such an agreement could have the following elements:

- A requirement that the parties authorize their vessels to conduct fishing in the area only after establishing one or more regional or subregional fisheries-management organizations to manage fishing in accordance with modern international standards
- Promotion of cooperative scientific research to improve the understanding of the ecosystem(s) of the area, and to help determine whether fish stocks might exist in the area now or in the future that could support sustainable harvesting
- Promotion of coordinated monitoring of the area.

Both the U.S. Navy and Coast Guard are examining strategic requirements and capability analyses to operate in the Arctic, which could include the extended outer continental shelf. The Navy's 2011 Arctic Capabilities Based Analysis (CBA) acknowledges shifting summer sea ice; the value of the economic potential from oil, gas, and mineral development; the estimates of time and expense reduction from the use of an Arctic route; and the dwindling national-resource capabilities. Mission-criticality analysis was performed in the CBA that considers regional-security cooperation, conflict prevention, and sea control to all be in the possible environmental conditions with related time lines. This analysis appears to predict the future: Climatologist David Robinson of Rutgers University states that the Arctic could be ice-free well before 2050.[3] Countries such as China are offering trade deals and investing in resource development in Denmark (including its self-governing state of Greenland), Sweden, and Iceland.[4]

In the face of Hurricane Sandy, researchers acknowledged that "planet Earth is a dangerous place, where extreme events are commonplace and disasters are to be expected."[5] With extreme events, competition for natural resources, and the increased use of the Arctic come the need to have policy and procedures in place for disaster management so that mitigation, preparedness, and response-and-recovery plans can occur. National Geographic Society futurist Andrew Zolli has written about how, for a nation, a community, a people, or a system, resilience is critical in the face of disruption.[6] Because the Arctic Ocean feeds both economies and ecosystems, both require resilience for long-term sustainability. Without improved governance, our country risks having a disaster occur without the ability to respond or recover from it.

Notes

1. http://nsidc.org/arcticseaicenews/2012/09/arctic-sea-ice-extent-settles-at -record-seasonal-minimum/.

2. George Willis Botsford, *Hellenic History* (New York: Macmillan, 1956).

3. Chip Reid, "Extreme weather could come with record Arctic Ocean ice melt," *CBS News Interactive*, 20 September 2012, www.cbsnews.com.

4. Elizabeth Rosenthal, "Race Is on as Ice Melt Reveals Arctic Treasures," *The New York Times*, 18 September 2012.

5. Roger Pielke Jr., "Hurricanes and Human Choice," *The Wall Street Journal*, 1 November 2012.

6. Andrew Zolli, *Resilience: Why Things Bounce Back* (New York: Free Press, 2012).

8 "GEOPOLITICAL ICEBERGS"

Dr. David P. Auerswald

The changing Arctic environment and increased naval operations and
commercial activity are exposing new dangers and scenarios that previously received little attention. David P. Auerswald presents three
realistic scenarios that nations and navies operating in the Arctic
might face. He notes that, regardless of the scenario presented, each
country in the region has a rather different interpretation of how the
issues should best be resolved. Additionally, Arctic nations are not
the only countries pursuing security and economic interests in the
region. Auerswald discusses five unclassified documents that describe
U.S. strategy for the Arctic and argues that collectively they present
a strategically and decidedly ambiguous message that is difficult to
understand and even more challenging to implement.

"GEOPOLITICAL ICEBERGS"

By Dr. David P. Auerswald, U.S. Naval Institute *Proceedings*
(December 2015): 18–23.

This past April, the United States assumed the chairmanship of the Arctic Council, the primary intergovernmental organization charged with coordinating

environmental and maritime issues in the Arctic. Since its creation in 1996, the Council has fostered relatively calm international discourse on shared regional concerns. It has succeeded. Arctic Council discussions have led to scientific cooperation in the region and agreements on both search-and-rescue efforts and preventing or at least minimizing environmental disasters—like the massive 2009 oil spill in the Gulf of Mexico—in the Arctic region.

To use a polar metaphor, the Arctic Council has been a pristine ice mass floating above the waves. Underneath that cooperative veneer, however, lies a series of potentially dangerous geopolitical icebergs that could severely damage unwary ships of state. And the United States is plying those waters without a clear sense of where those hazards lie and the belief that they pose no threat.

Scenarios and Threat Perceptions

The Arctic geopolitical situation is changing rapidly. The Arctic is warming faster than the rest of the globe, with the region losing roughly 13 percent of its ice per decade. Melting ice has raised the possibility of access to new transshipment routes, fishing grounds, mineral deposits, vast petroleum reserves, and tourism opportunities. Use of the Northwest Passage across Canada or the Northern Sea Route across Russia could shorten transcontinental shipment distances by at least a third and open new venues for destination tourism. A truly transpolar route may be possible by 2030, which would shorten shipping times even more. On the energy front, the harsh environment, sanctions on Russia, and recent low oil prices have put a temporary halt to Arctic energy extraction, but the long-term potential remains. Thirteen percent of the world's undiscovered oil and 30 percent of the world's undiscovered natural gas could lie in the Arctic, to say nothing of huge ore and rare-earth deposits on Greenland and in Arctic waters.

Balancing these opportunities are emerging flashpoints, and not just due to Russia's growing proclivity to challenge the international order through its remilitarization of its northern coast, its actions in Ukraine, or its infringements on Nordic and Baltic states' airspace and territorial waters. Consider the following scenarios, each of which represents a type of geopolitical iceberg.

Scenario 1: Geopolitical tensions and resource extraction. Global tensions and increased demand lead to rising oil prices. As the price rises, Rosneft, Russia's largest state-owned oil and gas company, moves equipment to the waters off Svalbard archipelago to drill test wells. The 1920 Spitsbergen Treaty identified Svalbard as Norwegian territory, but granted all treaty signatories, including Russia, the right to engage in commercial activities and resource extraction on the islands. The Russians argue that the treaty also applies to the surrounding waters, giving them the right to drill and extract any hydrocarbons they find.

Norway objects and moves to intercept the Rosneft vessels. Russia announces a large-scale "snap" military exercise involving multiple Russian vessels, aircraft, cruise missile batteries, air-defense systems, and ground-troop movements in areas from Murmansk to the Baltics. Those exercises put elements of the Russian North Fleet on a collision course with their Norwegian counterparts, while also demonstrating that most of Northern Europe is within range of Russian aircraft, air defenses, and cruise missiles. Most affected countries remain silent as the crisis in Svalbard unfolds.

Scenario 2: Economic influence with security implications. A Chinese mining company convinces the local government in Nuuk, Greenland, to allow rare-earth mineral extraction in areas uncovered by melting ice. The mining company brings in 5,000 Chinese workers and promises up to $3 billion in infrastructure improvements as well as annual payments of $1,000 to each of Greenland's 58,000 residents. Greenland pushes for accelerated independence from Denmark due to this newfound economic independence. The Danes object but allow a referendum, which passes with a large majority.

The Chinese convince the newly independent Nuuk government to stay out of NATO and the European Union by playing on fears of post-colonial interference in Greenland's internal business. Greenland subsequently demands that the United States pay $10 million annually for use of Thule Air Base or leave the island. The Nuuk government also announces that China is building a deep-water port in western Greenland for Chinese naval vessels, including electronic-surveillance ships that will monitor the U.S. East Coast. Within months it becomes apparent that Chinese mining is creating an environmental disaster in Greenland, regional search-and-rescue cooperation is quickly degrading, and Chinese fishing vessels are engaging in widespread and potentially illegal harvests in the North Atlantic and the Greenland Sea. Denmark claims this is no longer its problem. Iceland remains silent. Canada, Norway, and the United States protest Chinese actions and must decide whether to confront fishing trawlers backed by Chinese naval vessels.

Scenario 3: Coordinating disaster response. A cruise ship with 1,000 passengers and crew, mostly from North America, travels west to east across the Northwest Passage in the summer. The ship has neither an ice-hardened hull nor an icebreaker escort. The voyage proceeds without incident until the ship reaches the Fram Strait between Greenland and Svalbard, when it is driven into an ice field during a storm. The hull is breached and the ship begins taking on water. There are no surface vessels in the vicinity. Rescue helicopters from Iceland and Svalbard cannot take more than a handful of people to safety. Hundreds perish trying to board lifeboats, and hundreds more die of exposure in the days before the storm breaks and surface vessels can pick up survivors. Recriminations ring out between governments as to who bore responsibility for the disaster. The United States, Canada, Denmark, and Iceland pledge to acquire more and better search-and-rescue assets, but those promises get lost in domestic budget debates.

Threats *and* Opportunities

These examples are similar to actual planning exercises discussed in the region. One can take issue with the specifics in any one scenario, but they all point to an underlying problem in the Arctic: Countries in the region have very different interpretations of the geopolitical landscape, particularly with regard to hard-power issues like security and economics. (Eight nations possess territory, or territorial waters, above the Arctic Circle: Canada, Denmark via Greenland, Finland, Iceland, Norway, Russia, Sweden, and the United States.)

Take threats: The Canadians fear losing control over the Northwest Passage, which they see as their sovereign territory. Denmark is worried about Chinese influence over an increasingly independent and accessible Greenland.

Iceland desperately fears being ignored by the other Arctic powers. Norway's largest security threats are from possible Russian demands similar to those in Scenario 1, or from an environmental disaster that cripples fisheries in the North, Norwegian, or Barents Seas. Sweden and Finland both worry about a militarily and politically assertive Russia and the possible spillover from a broader East-West dispute over Ukraine, Russia's energy politics, or a clash in the Baltic Sea. The Russians have signaled that NATO encroachment into Russia's traditional sphere of influence is their most significant external threat. In short, there is a patchwork of threat perceptions among Arctic powers. In a crisis or emergency, differences in threat perceptions—whether or not they are true—could undermine regional cooperation or contribute to unintended conflicts.

Opportunities, particularly on the economic front, also vary across the region. Norway, Russia, and to a lesser extent Iceland have the most to gain in terms of hydrocarbon and fisheries extraction and an interest in defining best practices to their own advantage. The United States, Canada, and Denmark (via Greenland) have less to gain from near-term resource extraction given the greater ice coverage in the North American portion of the Arctic, and more to gain from stringent codes of conduct for the various extractive industries. Sweden and Finland have little obvious, new economic opportunities associated with their Arctic territories, principally because they are not Arctic littoral states. China and other powerful, non-Arctic states have signaled that they want a role in polar governance and resource extraction. Creating region-wide standards and extraction quotas—on fisheries, for instance—could be difficult when short-term interests conflict across the region, or when the short-term interests of some states undermine the long-term interests of others.

This patchwork of threats, opportunities, and, yes, power has resulted in Arctic countries emphasizing different international forums depending on specific national goals. On security issues, the United States, Norway, Denmark, and Iceland have emphasized NATO. Canada and Russia have emphasized unilateralism. Sweden and Finland have emphasized the United Nations and the European Union. On issues associated with regional governance and stewardship, less powerful Arctic states have emphasized broad multilateralism, principally by adding permanent observers to the Arctic Council and working through the U.N. Convention on the Law of the Sea (UNCLOS) and International Maritime Organization. The five Arctic littoral states (Canada, Denmark, Norway, Russia, and the United States) have emphasized their own abilities to manage the region without outside interference; a form of mini-multilateralism best expressed through the 2008 *Ilulissat Declaration.* Iceland responded by creating the Arctic Circle, an organization through which interested nations, indigenous peoples, and nongovernmental advocacy groups can discuss Arctic issues, an initiative met with disdain by some Arctic nations but embraced by China and others. When should the United States engage in multilateralism, bilateral relations, or unilateral actions in a region where the appropriate negotiating forum continually shifts depending on the issue and countries involved?

In short, the number of geopolitical icebergs is growing. The region has a cacophony of overlapping and diverging interests and perceptions, power capabilities, and negotiating forums. And that begs the question as to how the United States can navigate those geopolitical waters without hitting an Arctic iceberg.

U.S. Policy: Decidedly Ambiguous?

U.S. strategy for the Arctic is described in five main unclassified documents. These include the 2013 *National Strategy for the Arctic Region* and the companion 2014 *Implementation Plan for the National Strategy,* the Department of Defense's 2013 *Arctic Strategy,* the 2014 *Navy Arctic Roadmap,* and the 2015 agenda for the U.S. chairmanship of the Arctic Council.

U.S. strategy sets out very broad goals. The current National Strategy for the Arctic Region lists three: advancing U.S. security interests, pursuing responsible Arctic regional stewardship, and strengthening international cooperation. The DOD's Arctic Strategy lists three security-related subordinate goals: a secure and stable region, protection of the U.S. homeland, and international cooperation to address regional challenges. Neither strategy provides much detail on what these terms mean, what actions will be taken to achieve these goals, or how progress will be measured.

Supporters of current strategy would argue that such ambiguity is appropriate given the rapidly changing geostrategic circumstances in the Arctic. Russian behavior could change. Climate change could accelerate or decelerate. An alternative to hydrocarbon power could be discovered. Constructive ambiguity gives decision-makers the leeway to change specific policies as circumstances warrant, without appearing to contradict previous strategy or overcommitting resources.

That last point is important. Current U.S. strategy avoids expensive resource decisions. The 2014 *Implementation Plan,* for example, calls for a "maritime trend analysis and an infrastructure prioritization framework" when discussing increased Arctic maritime activity, rather than building deep-water ports or stationing more Coast Guard vessels in the Arctic. This is a prudent step if one believes that shipping and tourism along the Alaskan coast will not develop quickly.

The one place where the United States has been more specific is with regard to U.S. goals for its chairmanship of the Arctic Council. U.S. goals of enhancing ocean safety and stewardship, improving economic and living conditions for Arctic peoples, and addressing the effects of climate change in the region are well-established areas of international cooperation that build on past Arctic Council discussions and existing international agreements. Those goals also focus on the creation of international standards and norms of behavior, which can bring order to confusing international circumstances and shift the costs of maintaining such order onto end users (i.e., the shipping and extractive industries) or involve relatively low-cost research collaboration.

Three criticisms provide a less sanguine view of current U.S. Arctic strategy. First, existing strategy assumes a generally cooperative atmosphere among regional players. If this assumption is wrong, U.S. strategy may be inadequate to the challenges confronting the region. What happens if Russia continues on a revanchist path? What happens if China gains political influence over Greenland? What happens if an environmental disaster or resource shortage leads to beggar-thy-neighbor policies across the region?

A second criticism is that U.S. goals are mutually unachievable given today's Arctic geostrategic environment. The first and third goals in the *National Strategy* are particularly problematic. That is, it may be impossible to advance U.S. security while also cooperating with Russia unless Russian behavior changes for the better, or both the United States and Russia (and other Arctic nations) can continue to compartmentalize issues in the face of additional Russian provocations.

A third and more damning criticism is that U.S. strategy erroneously assumes that achieving U.S. goals is possible without investing in more than token Arctic capabilities. Unfortunately, the United States lacks the capabilities to "advance U.S. security interests" in the Arctic, the first goal in the *National Strategy*. The U.S. Coast Guard has *one* working icebreaker to cover the full 6,640 miles of Alaskan coast. The U.S. Navy has no ice-capable surface ships and has yet to retrofit existing ships with reinforced hulls, despite promising to "remain prepared to operate in the Arctic region" in the *Navy Arctic Roadmap*. In the Navy's words, "The Navy's existing Arctic Region posture remains appropriate," but "the Navy's surface and air forces have limited operational experience in the region and naval forces without specialized equipment and operational experience face substantial impediments." Overall, the United States has only limited intelligence, maritime awareness, and communications in far northern latitudes, to say nothing of adequate ports or the infrastructure with which to respond to a disaster.

This may explain why the DOD's *Arctic Strategy* says that the United States should be prepared to respond to a wide range of challenges and contingencies, but then says little about the capabilities or actions needed to achieve adequate preparedness. The DOD *Arctic Strategy*, like the aforementioned *Implementation Plan* and *Navy Arctic Roadmap*, calls for more study of U.S. needs, "comprehensive engagement with allies and partners," and support for civil authorities rather than pushing for Arctic-oriented acquisition.

A New Direction in the Arctic?

All the creative ambiguity in the world cannot mask the serious problems with current U.S. Arctic strategy. It suffers from questionable assumptions, fails to de-conflict goals, and is a poster-child for an ends-means mismatch. Whoever succeeds the Obama administration should step back and engage in fresh strategic thinking or risk being blindsided by fast-paced Arctic developments. That means reassessing the Arctic region, developing appropriate U.S. goals for that region, and then providing adequate capabilities to achieve those goals.

One persistent issue with U.S. analysis is that Americans tend to think of the Arctic, if they think about it at all, as comprising Alaska and its territorial waters. Full stop. Yet the United States is a global power, with interests and commitments (some specified in binding treaties) to the North Atlantic and European portions of the Arctic, to say nothing of protecting the global commons when it comes to freedom of navigation, the sanctity of exclusive economic zones, and environmental stewardship. Upholding those formal and informal commitments requires that the United States think broadly about the Arctic, have a strategy for that broad spectrum of commitments, and develop the requisite capabilities to match those commitments. So the first task is to rethink what we mean by the Arctic.

The next administration then should reassess U.S. goals to accurately and directly correspond to emerging threats, opportunities, and alignments that are developing across the region. One possible way to recast U.S. goals might be to first prevent either Russia or China from dominating the region in terms of economics or security. Regional participation by both countries is inevitable (and, one could argue, desirable), particularly with regard to Chinese investment. Dominance by either power, however, would undercut U.S. influence and commitments and put at risk U.S. interests in protecting the global commons. Coordination of relevant security efforts would take place through NATO. Economic efforts could be coordinated either bilaterally or through a future Transatlantic Trade and Investment Partnership (TTIP) agreement.

A second goal could be preventing an environmental disaster in the region. This requires that existing cooperation continue on shipping protocols, fisheries management, and oil-spill prevention and response, something in the interests of all Arctic littoral states. The appropriate venue for those discussions is the Arctic Council, with any agreements flowing from those discussions being codified as treaties.

A third goal could be fostering responsible private-sector investment in the region. Specific actions here could include providing U.S. loan guarantees, tax incentives, or access to government climate and geological data in exchange for private-sector creation of needed regional infrastructure. If extensive enough, private-sector investment could help ensure that neither Russia nor China achieves economic dominance in the region.

These goals better de-conflict the myriad crosscutting priorities, threats, and opportunities of the Arctic nations. The first goal aligns the United States with every Arctic nation but Russia and is just the sort of assurance that Nordic states (and their Baltic neighbors) have been looking for from the U.S. government. The second goal focuses on the environmental concerns of the Arctic littoral states and their economic self-interest. Even Russia, with its less-than-stellar environmental record, has an interest in maintaining fish stocks, and the Western-based oil companies that Russia will need to extract oil and gas from the Barents and Kara Seas have the reputational and fiduciary need to engage in relatively careful extraction in Arctic waters. The third goal is attractive across the region, but particularly with regard to Canada, Denmark, and Iceland, each of which wants more investment in its Arctic territories.

Adopting these goals would require that the United States invest *now* in additional civilian and military capabilities. Acting now is crucial given the long lead-time needed to build and field appropriate capabilities. The three greatest priorities are in sensors, communications, and surface ships. Sensors are necessary for even rudimentary maritime-domain awareness; knowing who is traveling where in the region and for what purpose. The United States needs better civilian capabilities in this regard to regulate shipping and avoid maritime accidents, including oil spills. Better civilian and military communications are needed for everything from coordinating search-and-rescue to managing sea and air traffic. Communications are particularly challenging given the lack of radio or cellular infrastructure in the region, and the mismatch between high northern latitudes and the orbital paths of most communications and geopositioning satellites. Finally, and perhaps most importantly, the United States needs to improve its naval capabilities to demonstrate a maritime presence across the region. At the least that will require more Coast Guard icebreakers and ice-capable Navy surface ships.

These are not trivial acquisitions. A heavy icebreaker, for instance, can cost $1 billion, and the U.S. Coast Guard has argued that it needs three new heavy and three medium icebreakers for northern-latitude operations. But absent these investments, the United States will lack the capability to monitor, communicate, and operate in this harsh environment. The United States needs to dedicate additional funding for this effort or reprioritize money within existing accounts, raise revenue through user fees or energy taxes, or share costs with the private sector in exchange for shared use of acquired assets. U.S. goals cannot be met in any real sense without these capabilities.

"Increasingly Dynamic and Important"

Skeptics will argue that taking these actions will militarize the Arctic, jeopardize international cooperation in the region, and create a security dilemma vis-á-vis Russia. This belies the fact that the region has *always* been militarized and will continue to be so for the foreseeable future. All Arctic states, with the exception of Iceland, have military installations above the Arctic Circle. The Norwegian Permanent Joint Military Headquarters, for example, is located above the Arctic Circle. Russia is refurbishing military facilities along its north coast and around the Kola Peninsula. The United States regularly sends submarines and long-range military flights through the region, maintains Thule Air Base in Greenland, and military aircraft in Alaska. So the notion that the Arctic is one large demilitarized zone is a misnomer.

Current events may also have made this criticism a moot point. Russian President Vladimir Putin is in the process of remilitarizing northern Russia as fast as he can, both to defend against existing U.S. capabilities and to reestablish a sphere of influence that Putin has always asserted was Russia's due. So far there has been no real Western response. Steps to counterbalance Russian revanchist policies, including better Arctic capabilities, might actually slow Putin. At worst, Putin will continue doing what he has always planned on doing.

The question to ask is thus not whether improved U.S. capabilities will create a security dilemma with Russia, but instead whether the United States can afford to do nothing in the face of ongoing Russian redeployments to the region. The Obama administration's recently released 2015 *National Security Strategy* argues, "We will deter Russian aggression, remain alert to its strategic capabilities, and help our allies and partners resist Russian coercion over the long term, if necessary." The key will be matching this rhetoric with real capabilities in the Arctic.

For the past 25 years, the Arctic has been a relatively frozen international backwater, moving at a glacier's pace. As the Arctic ice melts, however, international politics is seeping back into the region and creating new geopolitical icebergs. Supporters of current U.S. strategy say it is flexible, adaptable, and appropriate for today's circumstances. Yet U.S. strategy is based on questionable assumptions, inconsistent goals, and inadequate capabilities. The next administration should reassess the Arctic region, develop appropriate goals based on that assessment, and provide adequate capabilities to achieve those goals. None of this will be easy, particularly with regard to acquiring the capabilities needed to operate in the region. Inaction, however, risks being caught unprepared in an increasingly dynamic and important part of the world.

9 "COLD HORIZONS: ARCTIC MARITIME SECURITY CHALLENGES"

CDR John Patch, USN (Ret.)

In addition to the significant challenges of operating in the region, among them temperatures as low as –55 degrees Celsius, vast uninhabited expanses, months of darkness, communication difficulties, and logistics problems, some strategic concerns in the Arctic will affect naval operations. Shifting military postures and security stances by such nations as Russia and Canada require concerted U.S. attention. For Russia and Canada, specifically, Commander John Patch contends that although the challenges are real, Russia has more bark than bite and Canada represents a "strategy-resource mismatch." However, the United States is at the point at which it must make a decision to upgrade its Arctic policies, strategy, and capabilities. Although U.S. Navy undersea capabilities are unmatched, the Navy has no surface ships with hulls designed for Arctic operations. Patch contends that all this can be remedied, but "the United States is currently unprepared to defend the full array of Arctic interests."

"COLD HORIZONS: ARCTIC MARITIME SECURITY CHALLENGES"

By CDR John Patch, USN (Ret.), U.S. Naval Institute *Proceedings* (May 2009): 48–54.

Northern hemisphere maritime powers are shifting their focus to waters above 66°30′ N latitude. The 2 August 2007 planting of a titanium Russian flag on the North Pole seabed by a Mir deep-water submersible supported by the nuclear-powered icebreaker *Rossiya* and research ship *Akademik Federov* generated significant global media coverage and a groundswell of domestic Russian pride. While not a military mission, this event triggered frantic Arctic security interest reassessments. Recent attention on the Arctic, however, has typically covered climatic changes, natural resources, sovereignty claims, and new shipping lanes.[1]

Little discussed, however, are Arctic security challenges and regional military posture shifts. The larger Arctic states—namely, the United States, Canada, and Russia—are clarifying national positions, increasing Arctic operations, and trumpeting plans to expand military (largely maritime) capacity. Yet employing forces in the Arctic is far easier said than done; the harsh environment makes military operations expensive and dangerous. Further, modifying policy and strategy are relatively easy endeavors compared with adjusting force structure. For now, Arctic capabilities of northern nations do not match the strength of their ambitions. Nevertheless, the high stakes suggest this could soon change.

Marginal Maritime Capability in the Arctic

Operational challenges are legion: from temperatures below –55 degrees Celsius, to vast, uninhabited expanses, to months-long darkness. The Arctic environment—essentially maritime in nature—precludes most of the usual support militaries take for granted in more moderate or overland climes, such as navigation aids, communications, logistics and maintenance infrastructure, and even search and rescue (SAR) services. Communications, never perfect even in ideal conditions, are extremely limited. Satellite connectivity is rare, GPS coverage is marginal, and long-haul high-frequency communications are unreliable.

Even gathering and maintaining a basic awareness of other ships, submarines, and aircraft in the Arctic is almost impossible. With limited sensor coverage, rare sustained presence, some of the worst weather on the planet, and a bleak picture from horizon to horizon, it is no wonder Arctic nations have little idea what is transpiring to the north. A senior U.S. Coast Guard officer from the 17th District (Alaska) commented in 2008 on recent operations: "We had almost no idea, no maritime domain awareness, of what was actually happening on the waters of the Arctic."[2]

Aircraft are relatively unconstrained by Arctic surface conditions, but they routinely operate at the edge of the safety envelope. Aviators often fly in areas with little to no SAR coverage, greatly increasing risks associated with aircraft emergencies. Air crews might enjoy relative warmth, but fuel in unheated tanks begins to turn to slush at extreme cold Arctic temperatures.

Surface vessels are especially challenged in the Arctic. Free-drifting icebergs and shifting pack ice can hole all but the heartiest icebreakers. Twenty-four-hour winter darkness increases watch manning and saps morale. Moon-like cold threatens both humans and machinery; icing on ships' equipment and superstructures can pose a capsizing threat. Equipment and personnel casualties can quickly become emergencies in Arctic wastelands. Polar conditions transform sophisticated weapons and sensors into useless junk.[3] Arctic surface operations require herculean efforts just to remain within safety margins.

Nuclear-powered submarines, inherently self-sufficient, suffer least in the northern extremes. They rely on special, expensive technology to do so, including inertial navigation systems, ice avoidance sonar, and very low frequency communications. Still, ambient arctic ice noise makes tracking adversary submarines very difficult, and under-ice acoustic phenomena interfere with passive homing torpedo guidance.[4]

Russia: More Bark than Bite

Of late, Moscow's vows to defend Arctic interests have been the loudest. In September 2008, the Russian national security council began drafting new policy to formalize the claimed Arctic borders. After other Arctic countries criticized

Russia's bold claims and the United Nations deferred validating them, Russian military leaders announced new defense initiatives. Senior leaders declared they were adapting training plans for units "that might be called upon to fight in the Arctic."[5] This would increase the operational radius of Russia's northern submarine fleet and reinforce the Russian Army's combat readiness along the Arctic coast. Russian officials cite "large-scale U.S. armed forces' maneuvers" in Alaska (Exercise Northern Edge) as justification for increased Arctic military operations.[6]

Mistrust in NATO and suspicions toward U.S. hegemony form the basis of Moscow's Arctic propaganda, but Russia directs it at an internal audience as well. Hence, the Russian public generally supports the government view that Moscow needs more military capacity to enforce territorial claims. The roughly 25 percent annual increase in defense spending since 2006 claimed by Russian officials could provide more resources for the task.

Yet a look at Russian Arctic military operations reveals only moderate increases in activity. Some cite the upswing in Arctic long-range aviation (LRA) flights—Blackjack and Bear bombers—since 2007 as evidence of a resurgent Russian military. Moscow describes the flights' purpose as reconnaissance and practice for "our country's preparedness to stand up for its national interests and sovereign rights in the northern region."[7] NATO partners and the Canada-U.S. North American Aerospace Defense Command (NORAD) keep close tabs on the increased LRA activity (the Russian Air Force claims 50 global LRA flights as the recent single-day high).[8] But the partners have also been careful not to describe it as a threat.

It seems the flights' recent effects are more along the lines of strategic communication than strategic air power. Moscow has been quick to assure Washington that the LRA missions fly without nuclear weapons. Arctic LRA patrols had greatly diminished over the past 15 years, but never ceased entirely. Thus, the missions do not represent a new capability. LRA is clearly the cheapest and most visible way to achieve Arctic military presence.

To its credit, Moscow does field the strongest Arctic surface force. With the most heavy icebreakers and cold-weather ports, Russian seafarers are rather proficient in Arctic operations. Seven of Russia's 20 icebreakers are nuclear-powered, giving them unmatched power and endurance. The heaviest of them can manage six-plus feet of ice (winter sea ice can be more than twice that thick).

Russian authorities claim they will build six more nuclear-powered ice-breakers by 2010.[9] The Russian Navy is also resuming routine Arctic naval presence after a 17-year hiatus. In 2008, a *Udaloy*-class antisubmarine ship deployed for roughly a month, replaced by a *Slava*-class cruiser.[10] While a far cry from heady Cold War days, this demonstrates a limited commitment to deploy more surface units in Arctic waters.

Although information is limited, it appears that Russian subsurface Arctic operations are less frequent or vigorous. Two *Typhoon*-class ballistic-missile submarines (SSBN; one in reserve status) remain operational.[11] Moscow specifically designed these massive hulls for under-ice sea-launched ballistic-missile (SLBM) patrols. Yet, with the few remaining *Delta IV*s, only 15 Russian SSBNs remain from a Cold War high of 60, and material conditions are questionable.[12]

The long delayed follow-on SSBN *Borei* class and several *Bulava* SLBM launch failures suggest the Russian Navy may need a massive cash infusion to remain viable under the Arctic Ocean.[13] The recent petrodollar decline will compound pervasive Russian Navy training, maintenance, and logistics short-comings, annual defense budget increases notwithstanding. For now, though, Moscow stands atop the heap in most areas of Arctic military wherewithal and expertise.[14]

Canada: Strategy-Resource Mismatch

Ottawa lacks the ability to support its ambitious, contemporary goal to assert control over northern territories. Longstanding territorial disputes with America and other nations failed to generate enough interest to build up Arctic military capacity, but after the 2007 Russian flag planting, Canada expressed the intent to support its Arctic sovereignty with force. Prime Minister Stephen Harper stated that two new military facilities would "tell the world that Canada has a real, growing, long-term presence in the Arctic." Harper further announced that Canada would spend $5.7 billion to build six to eight new navy patrol ships to guard the Northwest Passage, with the first arriving by 2013.[15]

The new Arctic bases include a Resolute Bay army training center for cold-weather fighting and a deep-water refueling port on the northern tip of Baffin Island, both to extend Arctic operational reach and endurance. Other efforts include plans for underwater, land, surface, and air sensors, including unmanned aerial vehicles to monitor the north. Finally, Ottawa will design follow-on destroyer and frigate classes with the capability to operate in limited ice conditions.

Ottawa's Arctic security stance is not new, however. Canadian and U.S. strategic interests have long coincided, serving as the impetus for NORAD's creation as a bulwark to the Soviet threat to North America in the form of Arctic submarine-launched ballistic missiles and air-launched nuclear-tipped cruise missiles.

Current Canadian capabilities, however, are simply insufficient to patrol or safeguard the Arctic expanses. Ottawa acknowledged this shortcoming in its latest Arctic-focused *International Policy Statement* and associated *Defence White Paper*, which relate the need to move beyond simple words to build an Arctic military capacity. Currently, periodic Aurora (P-3) maritime patrol aircraft missions and radar satellite coverage provide a degree of Arctic surveillance, but sustained Canadian Arctic military presence is rare.

Canada has one large and five medium icebreakers, most nearing the end of their service lives. After a 22-year hiatus, Canadian warships resumed a limited Arctic presence with annual (summer) deployments in 2004.[16] Canada once envisioned nuclear-powered submarines for Arctic operations, but it has no under-ice submarine capability. Alongside other unrelated, but expensive, military recapitalization efforts such as cruiser, destroyer, and submarine upgrades, Canadian Arctic forces face an uncertain future.

Canadian Arctic military exercises have, however, increased in scope and frequency. Since 2002, Canadian Forces (CF) have conducted a new series of large-scale joint exercises, most termed as "sovereignty operations," with a heavy emphasis on support to civil authorities. Canada Command's (U.S. Northern Command equivalent) Joint Task Force North conducted the Operation Narwhal exercise series in increasing scope since then. Canada also routinely conducts NORAD Arctic air-defense exercises, usually involving CF-18 sorties to monitor Russian LRA activity. Recurring CF deployments will surely improve Canada's already significant northern operational expertise. However, these exercises reflect little ability to monitor or deal with adversary military forces in under-ice or frozen surface conditions.

America: Strategic Decision Point

As in Canada, recent high-visibility events focused U.S. attention on Arctic policy and strategy. Both were found wanting: the 15-year-old policy did not reflect current strategic realities, and the current National Security Strategy, National Defense Strategy, and maritime strategies provide only a single scant reference to Arctic security interests among them.

In 2008, the State Department and National Security Council conducted an in-depth review of the 1994 Presidential Decision Directive on Arctic Policy. A new January 2009 policy lists key Arctic interests: missile-defense and early-warning, strategic sealift, strategic deterrence, freedom of navigation/over-flight, maritime presence, and maritime security operations.[17] All reflect a clear need for unfettered Arctic access (though the last two items are arguably not interests), with "strategic sealift" the standout term not often associated with the Arctic. Yet the new administration will have to adopt the policy to put it into force.

The January 2009 Arctic policy contains only general guidance for Defense Department implementation. This is an important distinction, as future U.S. Arctic force structure will, to a degree, be justified with any new policy. It directs Defense (among other departments and agencies) to develop greater capabilities to protect Arctic borders; increase Arctic maritime domain awareness to protect commerce, critical infrastructure and resources; preserve global mobility and presence; and encourage peaceful dispute resolution. This creates a net increase in Arctic security missions but does so under a cloud of shrinking defense and homeland security budgets.

Current U.S. Arctic forces are insufficient to accomplish these new missions, much less standing tasks. Arctic air operations, for instance, lack persistence and reach. With little justification for constant Arctic patrols, flights are usually associated with brief exercises or scientific expeditions. P-3C maritime patrol aircraft rarely venture north, but demand is increasing even as the number of airframes shrinks.

Coast Guard aviators, as a rule, operate more frequently in the Arctic than their Navy brethren. Bering Sea C-130 patrols occur periodically out of Kodiak, Alaska, but lack of infrastructure, special Arctic maintenance requirements, and airframe and aircrew shortages limit missions. Unlike the Navy, however, the Coast Guard is on call 24/7 for Arctic SAR and maritime incident response. Ongoing phasing-in of the J model C-130 will somewhat improve Coast Guard Arctic air capability (improved endurance, range, and anti-icing). The Navy will fare worse without its own dedicated C-130Js; the P-8A offers a step backward from the P-3C in Arctic capability.

America also has a comparatively poor Arctic surface force. Much has already been written on the marginal U.S. icebreaker capability—one light and two heavy icebreakers—able to handle only five or so feet of ice. The Coast Guard claims that it will take more than $800 million and perhaps ten years to either build a new ice breaker or extend the service life of the existing fleet.[18] A 2007 National Research Council report captured the contemporary surface capability assessment: "U.S. icebreaking capability is now at risk of being unable to support national interests in the north and the south."[19]

The Navy also has no hulls designed for Arctic operations and only very limited institutional expertise at surface operations in polar climates. A 2001 Office of Naval Research–sponsored Arctic symposium report "revealed large gaps in knowledge and highlighted deficiencies in U.S. Navy and Coast Guard equipment and training for cold weather."[20] Perhaps acknowledging this, the Navy Secretary recently stated an ongoing study would assess future missions for surface ships in the far north. Similarly, the Coast Guard began a comprehensive "high-latitude study" in December 2008 to determine its requirements for meeting all of its 11 statutory missions.

Navy Arctic undersea capabilities, however, are unmatched. Because of long-standing undersea warfare requirements, the submarine force maintains three current classes—*Los Angeles,* improved *Los Angeles,* and *Seawolf*—with Arctic capabilities.[21] Annual polar surfacing events broadcast U.S. and combined U.S-British under-ice skills demonstrated during exercises.

Like Canada, the Homeland Security and Defense departments also increased the frequency and scope of Arctic exercises and operations in recent years. U.S. Northern Command created Joint Task Force Alaska in 2002, and coordination between DHS and the U.S. Pacific Command's sub-unified Alaska Command improved. Joint, interagency, and NORAD exercises with at least some Arctic focus include Northern Vigilance, Northern Denial, Northern Edge (the largest, with over 5,000 participants), Terminal Fury, and Valiant Shield.[22]

Annual trilateral (U.S.-Russia-Canada) SAR coordination also occurs. U.S.-U.K. Arctic Ice exercises are less frequent, but focus more on the potential Russian Arctic undersea threat. The Coast Guard also participates in the International Ice Patrol, monitoring southbound icebergs near Greenland. Yet the vast majority of forces assigned to these exercises and operations only operate briefly in the Arctic, if at all.

Few permanent U.S. forces remain above the Arctic Circle, and almost none focus solely on Arctic security. All the major year-round Alaskan military and Coast Guard bases are below the Circle. The Coast Guard's 17th District monitors the Arctic, but the distance from permanent bases to the northern slope in deep winter limits on-station time. The Coast Guard conducted a "proof of concept" forward-operating location at Barrow, Alaska, in the extreme north (320 miles north of the Arctic Circle) in recent years, but with only a helicopter and small boat detachment. As a recent Coast Guard officer stressed, "[we] are not positioned or prepared for regular Arctic operations."[23] The closest permanent military base to the Arctic, Naval Air Station Keflavik, Iceland, lowered the American flag for good in September 2006.

Bleak Future?

While the United States is currently unprepared to defend the full array of Arctic interests, this does not represent a national security crisis. With prudence and forethought, America can develop viable policy, strategy, and force structure to meet future challenges. Given the looming period of austere defense budgets, however, such efforts will require significant congressional and popular support.

In addition, though the Coast Guard needs no convincing, Navy leaders may resist the institutional costs of new Arctic missions. Further, attempts to manage Arctic interests with only the diplomatic, informational, and economic tools of national power may prove insufficient. Similarly, a unilateral approach to Arctic security will surely prove more costly and less effective over the long term. Arctic military confrontation is neither desirable nor inevitable, but trends clearly show a future of colliding Arctic state interests. Equal-parts cooperation and containment can keep the Russian Bear's appetite in check and vouchsafe U.S. Arctic interests.

The American Arctic View

- No comprehensive U.S. Arctic strategy exists.
- The next decade's Arctic debate will focus on environmental, regulatory, law enforcement, political, and energy issues, vice military capabilities.
- The only "new" U.S. Arctic security interest is strategic sealift.
- There is no current logic beyond undersea warfare for persistent U.S. naval presence in the Arctic Circle.
- NORAD capabilities will be sufficient to meet Russian Arctic air threats for the immediate future.
- No country will challenge U.S. undersea Arctic dominance in the near future.
- Canada's interests in benefiting from U.S. security guarantees far outweigh sovereignty sensitivities.
- Russian dominance of Arctic surface operations provides unique access and expertise.
- The U.S. Navy has no current imperative to conduct significant Arctic air or surface operations; the converse holds true for the Coast Guard.
- Traditional surface warfare is possible only on the Arctic fringes for part of each year.
- Countries exaggerate extant Arctic military capabilities for effect.
- Any near-term increases in Arctic military capability will be marginal.

National Security Recommendations

- Formulate a new Arctic security strategy—one that avoids unrealistic objectives; alternatively, integrate Arctic strategic guidance into the next National Security Strategy.

- Strengthen Canadian-U.S. Arctic defense ties, potentially with a new Arctic defense agreement under NORAD auspices; commit to a complementary U.S.-Canadian development, basing, and employment of Arctic security assets.

- Meet every Russian Arctic force presence with an equal, but nonconfrontational, Western response.

- Assign the Coast Guard as lead agency for Arctic security; fund sufficient ice breakers (with a minimal surface-warfare capability) and C-130Js to accomplish law enforcement, regulatory, search-and-rescue, and security missions.

- Design a light ice breaking capability into a portion of the next destroyer and military sealift classes.

- Develop a permanent deepwater maritime security facility on the northern slope of Alaska.

- Reinvigorate NATO's Arctic surface and air capabilities.

- Procure sufficient Navy C-130Js to augment Coast Guard Arctic patrols in wartime.

- Develop joint Arctic doctrine.

- Commence Arctic conditions testing of selected shipboard/aircraft weapon systems.

Notes

1. See LCDR Anthony Russell, USCG, "Carpe Diem Seizing Strategic Opportunity in the Arctic," *Joint Forces Quarterly* p. 51, 4th quarter 2008, and Peter Brookes, "Flashpoint: Polar politics: Arctic security heats up," *Armed Forces Journal,* November 2008, http://www.armedforcesjournal.com/2008/11/3754021.

2. U.S. Coast Guard, "Arctic Journal: Part 1," *Coast Guard Journal,* 7 April 2008, http://www.uscg.mil.usnwc.idm.oclc.org/cgjournal/message.asp?Id=65.

3. See "Naval operations in an Ice free Arctic," Office of Naval Research Symposium, 18 April 2001, http://www.natice.noaa.gov.uonwo.idm.oclc.org/icefree /FinalArcticReport.pdf.

4. U.S. Navy, *Undersea Warfare*, Spring 2005, vol. 7, no. 4, http://www.navy.mil /navydata/cno/n87/usw/issue%5f27/asw2.html.

5. Bradley Cook, "Russian Army Trains for Arctic Combat to Defend Resource Claim," *Bloomberg.com*, 24 June 2008.

6. "Russia to take into account US army high activity in Arctic," *ITAR-TASS*, Moscow, 5 May 2008.

7. "Russian Northern Fleet aircraft fly over the Arctic," *Interfax-AVN*, 4 September 2008.

8. "Cool your jets, Russia!" *Toronto Sun*, 9 September 2007.

9. "Russia to Build Six Nuclear Icebreakers for Arctic Transport," *ITAR-TASS*, 15 January 2009.

10. "Russia Extends its Arctic Naval Power Base," http://www.defense-update .com/newscast/0808/070802%5frussian%5fnavy_in_the_arcrtic.html.

11. According to Global Security.org; see http://www.globalsecurity.org.usnwc .idm.oclc.org/wmd/world/russia/941.htm.

12. STRATFOR, "Russia: a Second Strike Capability Failure," 23 December 2008, http://www.stratfor.com.usnwc.idm.oclc.org/analysis/20081223_russia _second_strike_capability_failure.

13. Pavel Baev, "Russia's Race for the Arctic and the New Geopolitics of the North Pole," The Jamestown Foundation, *Occasional Papers*, October 2007, p. 8.

14. See Adele Buckley, "Establishing a Nuclear Weapon Free Zone in the Arctic," Canadian Pugwash Group, 11 July 2008, p. 4.

15. Tim Reid, "Arctic military bases signal new Cold War," *The Times*, 11 August 2007.

16. "Defence Requirements for Canada's Arctic," *Vimy Paper 2007*, The Conference of Defence Associations Institute, p. 17.

17. White House, "Arctic Region Policy," *National Security Presidential Directive 66/Homeland Security Presidential Direct 25*, 9 January 2009, http://www .whitehouse.gov/news/releases/2009/01/20090112–3.html.

18. Ronald O'Rourke, "Coast Guard Polar Icebreaker Modernization," Congressional Research Service, 3 October 2008.

19. National Research Council, *Polar Icebreakers in a Changing World: An Assessment of U.S. Needs*, Washington, 2007, p. 2.

20. ONR Research Symposium.

21. J. L. Gossett, "A New Era in the Arctic," *Undersea Warfare*, Issue 12, 2001, http://www.navy.mil/navydata/cno/n87/usw/issue_12/era_the_arctic.html.

22. U.S. Air Force, Northern Edge Background Paper, Elmendorf Air Force Base, http://www.elmendorf.af.mil/library/factsheets/factsheet.asp?id=11648.

23. Russell, "Carpe Diem . . . ," p. 98.

10 "THE COLD, HARD REALITIES OF ARCTIC SHIPPING"

Stephen M. Carmel

Because of melting sea ice, within two decades it may be possible to transit the Arctic in midsummer and see no ice. However, that is not possible at present. Stephen M. Carmel presents the case that the United States is unprepared to deal with pending increases in maritime traffic. He recommends several steps for formulating an appropriate strategy that intelligently expends the scarce and uneven national investment in resources for U.S. naval and maritime operations. To be sure, the Arctic is not the only region on the globe in need of strategic attention. However, it is important to U.S. national interests. He also reminds readers that different types of shipping, such as container shipping and bulk shipping, require different platforms and resources. Shorter routes for shipping are economically desirable, but the benefits must be weighed against the cost of building Arctic-capable ships. The Arctic must not be viewed in isolation from the rest of the globe, and solving the problem of Arctic shipping (and other operations) requires a complex calculation.

"THE COLD, HARD REALITIES OF ARCTIC SHIPPING"

By Stephen M. Carmel, U.S. Naval Institute *Proceedings* (July 2013): 38–41.

On 16 September 2012 the Arctic reached the point at which ice stops receding and begins to form anew with the approach of winter. Last year [2012] that ice minimum set a record at 1.32 million square miles—300,000 square miles less than the previous record minimum.[1] With that news comes the predictable flood of reports about the pending increase in Arctic shipping, how woefully unprepared the United States is to deal with that onslaught of traffic, and the need for large-scale investment in Arctic capabilities.[2]

Worries about the implications of a thawing Arctic have been around for some time. Conferences and seminars about the Arctic seem to have superseded even piracy as a source of income for the conference-for-profit crowd. There is no doubt that the Arctic is in fact thawing, and there naturally will be increased activity up there. But to formulate appropriate strategy and make intelligent investments it is important to get past the hype and:

- Understand what type of activity is likely to occur
- Determine the time frame in which it is likely to happen
- Recognize that, at least for commercial interests, economics trumps all

As the Russians found out with the failed Shtokman gas field—a €15 billion (roughly $20 billion) Arctic investment killed due to cheap U.S. shale gas—the Arctic is not melting in isolation from events in the rest of the world.[3] It is the global system, of which the Arctic is just one part, that matters; changes across that system, including in the Arctic, interact in ways that can be unpredictable at best. It is very unlikely, then, that the Northern Sea Route across the top of Russia will become a major pathway for the global flow of commerce, and it is virtually certain the Northwest Passage across the top of Canada will never be useful for international trade.

Not All Shipping Is the Same

There are two types of shipping that must be considered when thinking about commercial traffic through the Arctic: destination shipping and transit shipping. Destination shipping is that which occurs to support some activity in the Arctic—oil moving from the Barents Sea across the Northern Sea Route to Asia, for example. That type of activity happens now and indeed will increase in volume. There are large amounts of natural resources in the Arctic, and while the economic viability of all of those discoveries is doubtful—as noted with Shtokman—many will be. Bulk shipping activity necessary to exploit those resources will increase. In addition, the Northern Sea Route offers the Chinese at least a partial solution to their "Malacca Problem," providing a source of oil and gas from the Barents Sea that cannot be interdicted, unlike that obtained from the Persian Gulf.

This type of destination shipping by definition means that such traffic will call at ports in at least one of just five countries having an Arctic coast (the United States, Canada, Russia, Denmark—via Greenland—and Norway). Using Port State Control (PSC), those littoral countries have considerable leeway and authority outside the painfully slow International Maritime Organization (IMO) process to implement the regulatory regime necessary to protect the environment and control shipping activity to an appreciable extent. Should all five of those states, perhaps through a sub-group of the Arctic Council, join to implement a coordinated PSC regime for access to Arctic ports, the bulk of commercial traffic there is de facto regulated.

Such shipping by its nature is also amenable to some of the challenges Arctic shipping presents. In particular those ships do not operate in networks, are not sensitive to variation in schedule, and have less sensitivity to adverse economies of scale. They also do not represent the volume of shipping a global pathway of commerce—a northern version of the Suez Canal, say—would represent. That sort of transit shipping, using the Arctic as a shortcut between Rotterdam and Yokohama for example, is far more uncertain.

Speed vs. Reliability

In predicting increased traffic through the Arctic it is often noted that routes across the top are up to 40 percent shorter than the more traditional routes between Asia and Europe (via the Suez Canal) or the East Coast of the United States (via the Panama Canal).[4] The assumption is that shorter equals faster and cheaper. But in the Arctic, the shortest distance is normally neither faster nor cheaper for the type of transit shipping usually associated with global commerce, particularly that involving containerships.

Container shipping is considerably different from bulk shipping, making the economics of the Arctic as a transit route unappealing. There are many things, such as construction standards, outfitting, and crew training for example, that make Arctic-capable ships more expensive to build and operate. In addition, those more expensive construction features are useful only during the short ice season but represent a cost the ship carries throughout the year. Other issues also make the Arctic a much more expensive place to operate, such as the need for icebreakers, lack of support infrastructure, and pending IMO requirements on fuel.[5] But to keep the discussion at a manageable level it is important to focus on a few key issues.

First, speed alone is no longer the major consideration, as fuel costs have made slow-steaming the standard of operation. Where once 24 knots was routine for a containership, it is now 13 knots or less. What is far more important than speed is reliability. Unlike the bulk shipping discussed earlier, schedule integrity is a key service-attribute for containerships. The Arctic will always suffer from periods of poor visibility and the potential for wind-driven ice, both of which can make routes with a comparatively low average transit time have a large variability around that average. More than half of all container cargo is now component-level goods—materials destined for factories for use in production processes operating on a just-in-time-type inventory-management system. That makes consistency, reliability, and schedule integrity of paramount importance. The key goal of container shipping is 99 percent on-time delivery. If this is attainable at all, it will be extraordinarily expensive using Arctic transit routes. Thus the variability in transit time that may be tolerable in bulk shipping is unacceptable for container shipping.

Networks and the Bottom Line

Containerships operate in networks with "strings" (routes) of many ports serviced by multiple ships on a steady schedule. For example, a U.S. East Coast to

Southwest Asia route of 42 days round trip serviced by six ships means regular weekly service out of the ports on that route. Routes frequently intersect at key transshipment ports such as Singapore or Algeciras, Spain. Network economics are a considerable part of the overall cost-efficiency picture in a container service. Transit across the Arctic, while shorter for certain port pairs, may not be shorter for a network that services a number of ports on both sides or call at a major transshipment hub. A requirement to call at Singapore for example, means the Northern Sea Route would not be shorter. Were the service to be restricted to just those ports where the distance is shorter, then all the economic advantages of network economics are lost. At the very most, the Arctic is serviceable just three to four months a year, and no one is predicting an ice-diminished Arctic in the winter. Developing routes that would increase the attractiveness of Arctic paths from a network perspective is not economically feasible as long as they are useful only a third of the year or less.

Perhaps the biggest issue making Arctic shipping unacceptable from a container-shipping perspective is economies of scale. While conventional wisdom would focus on total voyage cost, it is actually the cost per container that matters. Because both the Northern Sea Route and the Northwest Passage are draft-constrained (41-foot and 33-foot controlling drafts, respectively) the largest ship likely to be able to use the Northern Sea route would be one with a cargo capacity of just 2,500 TEU—and even smaller for the Northwest Passage. TEU, or twenty-foot equivalent unit, is a measure of containership carrying capacity based on a standard 20-foot container length. A 40-foot container would be 2 TEU, for example. The Northern Sea Route also has a beam restriction of 30 meters, as transiting ships cannot be wider than the icebreakers employed to support them. For the Asia-to-Europe trade on the other hand, containerships can be as large as 15,000 TEU with a beam exceeding 164 feet; 6,000 to 8,000 TEU ships are common.

As a back-of-the-envelope example, consider a voyage from Yokohama to Rotterdam, the common benchmark. By the traditional route it is 11,300 nautical miles (nm) with a transit time of 36 days. The Northern Sea Route is 7,600 nm and takes 26 days (relying on the rather large assumption that the voyage is unhindered by ice or visibility issues). The ship making the Arctic transit would reasonably be carrying 2,000 containers; the ship on the traditional route would be carrying 6,500. Factoring in all expenses such as fuel and daily ship operation, the cost of the traditional route would be $3.5 million, while the Arctic route would be $2.5 million.

That is as far as most analyses normally go, showing that the Arctic route is considerably cheaper. But as noted earlier, what matters is not total cost but cost-per-container—and when put in those terms it breaks down to $538 for the traditional route, but $1,250 on the Arctic route. So in fact, the Arctic route is more than twice as expensive as the traditional route, and the Arctic route looks worse when a comparison with larger ships on the traditional route is made. The Maersk Line, for example, will deploy Triple E–class ships with a nominal capacity of 18,000 TEU on the Asia-Europe trade route in the near future. It should be noted as well that the Northern Sea Route is actually a series of seas—Barents, Kara, and Laptev—connected by narrow straits and it is claimed by the Russians as an internal waterway. Fees to transit the Northern Sea Route are on par with the Suez Canal, and the Russians also impose a considerable and very formal administrative process in order to transit the route.

Keep the Bigger Picture in Mind

Lastly, it is important to remember, as noted at the outset, that changes in the Arctic are not occurring in isolation from the rest of the world—it is but part of a system, and the entire system is changing. When making ice projections out to 2040, then, it should be remembered that in a similar time span (roughly three decades) the advent of the container and advances in information technology completely revolutionized shipping—allowing the development of disaggregated supply chains that are the hallmark of this age of globalization, and propelling China from a third-world backwater to global economic powerhouse.

Clearly, a great deal can happen in 30 or 40 years, so it is a mistake to try to overlay a melting Arctic on today's geo-economic situation. It is the state of the world at that future point interacting with a melted Arctic that matters. Already, changes in the patterns of global trade have had significant implications for the utility of Arctic routes. Increasingly expensive labor in China, for example, is pushing Chinese manufacturing to be outsourced to countries in Southeast Asia where costs are lower but Arctic routes offer no advantage. A shift to near-shoring—moving manufacturing closer to markets—is increasing, too. Even advances such as additive manufacturing—3-D printing, for example—have large implications as local on-demand manufacturing becomes a reality. Advances in that type of disruptive technology could have a major impact on the fundamental nature of trade within the time projections of changing ice conditions in the Arctic.

There is no question that the Arctic is becoming more ice-free. There will be an attendant increase in commercial presence in the Arctic that should not be ignored. But a proper understanding of what type of activity there will be, and a realistic assessment of the volume of that activity are necessary to ensure proper policy and investments are made. For commercial shipping, and particularly the types that drive globalization today, Arctic routes do not now offer an attractive alternative to the more traditional maritime avenues, and are highly unlikely to do so in the future.

Notes

1. "Arctic Sea Ice Hits Smallest Extent in Satellite Era," NASA, 16 September 2012, www.nasa.gov/topics/earth/features/2012-seaicemin.html.

2. See for example James Holmes, "The Arctic is the Mediterranean of the 21st Century," *Foreign Policy*, 29 October 2012, www.foreignpolicy.com /articles/2012/10/29/open%5Fseas.

3. Terry Macalister, "Plug pulled on Russia's flagship Shtokman energy project," *The Guardian*, 29 August 2012, www.guardian.co.uk/world/2012/aug/29 /shtokman-russia-arctic-gas-shale.

4. Missing from the discussion is the fact that the Panama Canal actually is not the most popular route from Asia to the U.S. East Coast. It is preferable to offload on the West Coast and move cargo east by rail.

5. For ships sailing south of 60 degrees south, the IMO requires that no heavy fuel be on board, let alone burned. There is a similar rule pending for the Arctic.

11 "A MECHANISM FOR ARCTIC-CRISIS RESPONSE?"

Paul Arthur Berkman and ADM James G. Stavridis, USN (Ret.)

Receding sea ice in the Arctic is increasing opportunities for Arctic commerce and navigation, but also increasing operational risks. Paul Arthur Berkman and Admiral James G. Stavridis contend that such risks should create a sense of urgency for the United States to take the lead in such areas as search-and-rescue capability, environmental disaster mitigation, diplomacy, and scientific research. Balancing risks and opportunities is a challenge for all Arctic nations, and the responsibility for doing this is global as well as regional. Uncertain are the strength of the American national will for doing so, as well as the way forward.

"A MECHANISM FOR ARCTIC-CRISIS RESPONSE?"

By Paul Arthur Berkman and ADM James G. Stavridis, USN (Ret.),
U.S. Naval Institute *Proceedings* (December 2015): 36–37.

The High North is opening to the world. In April the United States began its two-year chairmanship of the Arctic Council, the high-level international forum for the region. President Barack Obama traveled to Alaska in August as the first U.S. President to visit the Arctic while in office. And the U.S. Coast Guard icebreaker USCGC *Healy* (WAGB-20) arrived at the North Pole on 8 September. All of these recent developments highlight the responsibilities of the United States as an Arctic nation.

The question is, can the United States provide international leadership in the Arctic, especially with regard to sustainable infrastructure development in the Arctic Ocean? And given our responsibilities, are we prepared to respond to disasters and fully participate in the High North—with search-and-rescue capability, environmental-disaster mitigation, science diplomacy, and other activities?

Unlike centuries past, when sea ice covered the north polar region perennially, today there is navigable open water from the Bering Strait to the Barents Sea during the summer. This increasing access to rich resources is awakening a number of human activities and associated societal responses, not just from the Arctic states but from the entire world. This leads directly to the hot-button topic of energy exploration, development, and production in the Arctic Ocean. Oceanic travel across the top of the Earth cuts a third off the distance between Europe and Asia, compared to transits through the Panama or Suez canals. What are the implications for new trade routes or trading patterns, which historically have changed the balance of power among nations? How will we use the Northern Sea Route, Northwest Passage, or Transpolar Route into the future?

Vast fishery enterprises are seeking to feed a hungry world, preparing to jump into areas of the Arctic high seas where marine living resources are unregulated beyond sovereign jurisdictions. Can nations collectively demonstrate shared stewardship and commercial restraint to ensure the lasting vitality of Arctic marine ecosystems?

Wrapped into charged dialogues about climate change, atmospheric temperatures over the Arctic are rising twice as fast as the rest of the Earth. Can we turn down the vitriol to appreciate that every planet in our solar system has its own changing climate, all influenced primarily by the Sun? The climate dynamics on Earth are no different, except that our planet is influenced by both natural variability as well as human impacts. A reality check is in order here.

On a global scale, we are just in our infancy in addressing climate and other planetary-scale impacts that require coordination among all nations.

So, where does this leave us in the Arctic? Can we conceive and build sustainable infrastructure in the Arctic Ocean that will resonate with utility and hope, not just for the region but globally?

In this quest, it is important to recognize that economic prosperity, environmental protection, social equity, and societal welfare all are necessary. We have responsibilities to act in the interests of present as well as future generations. Moreover, in the Arctic Ocean, as elsewhere on Earth, we have a shared struggle to balance national and common interests.

The challenge for the United States and the other Arctic states, with the central involvement of the indigenous peoples and effective engagement of non-Arctic states, is responding in a balanced manner to the opportunities as well as the risks from the opening of the Arctic Ocean.

We need to search for projects that can add value, inspire international cooperation, and improve the ability of humankind to operate responsibly in the High North. One such venture would be a multipurpose platform from which to conduct emergency responses, from search-and-rescue to pollutant cleanup or even critical vessel services. No such platform exists today, nor is one contemplated. Such an enterprise also could be used as a base for scientific research as well as observance and communication systems.

An appropriate-sized platform needs to be considered, but it could be roughly the dimensions of the larger oil-drilling rigs. It would need to be sustainable throughout the year, no matter the weather and environmental conditions, and possibly could be mobile. Like polar stations elsewhere, it would require year-round operations that could involve international, interagency, and private-sector crews. Such a platform would need to be easily accessible by air and sea. In order to minimize any complications with international law, it could be placed inside internationally recognized territorial waters, or at a minimum, in a nation's exclusive economic zone.

Where precisely should the platform be located? First, in terms of a best spot for emergency response, the Chukchi Sea is at the confluence of the Northern Sea Route, Northwest Passage, any transpolar route, and the Bering Strait region. Providing leadership, the United States would be in the position to assist with safety of life at sea as well as environmental-pollution responses throughout the region, including calls from Canada or Russia. Such international-response capacity recognizes that the marine system operates independently of any geopolitical boundaries.

The offshore area of the Chukchi Sea has water depths less than 1,300 feet, which is shallow enough to engineer and construct a multipurpose facility, yet deep enough to serve as a deepwater port more than 1,000 miles north of Dutch Harbor in the Aleutian Islands. Considerations of any such port along the Alaskan coastline have been problematic in terms of location and funding, especially recognizing that any federal contribution will be absent as long as Alaska continues to be the only ocean-front state in the United States without a coastal zone-management plan in effect.

It is clear from recent protests against Shell Oil Company, for example, that there is strong opposition and justifiable environmental concern about any hydrocarbon-extraction activities in the Arctic Ocean. At the same time, energy companies are planning three to five decades into the future to supply the fuel that we have come to demand on a daily basis, allowing us to warm our homes, run our computers, and travel across cities. How do we balance economic prosperity and environmental protection in the Arctic Ocean?

Building the platform in the Chukchi Sea would become a vital contribution for the 2011 Agreement on Cooperation on Aeronautical and Maritime Search and Rescue in the Arctic and the 2013 Agreement on Cooperation on Marine Oil Pollution, Preparedness and Response, both of which are lacking in infrastructure to become operational. Such a facility would enable the United States to demonstrate active and influential leadership in the Arctic, bringing a fresh focus on Arctic infrastructure. Moreover, in view of the previously mentioned international agreements, involving all eight Arctic states, such a facility would contribute to stability and peace in the region.

Importantly, construction of built infrastructure in the Arctic Ocean will be expensive, tapping precious national and state resources in directions that could compromise other priorities. At this stage, before offshore energy operations emerge, nations such as the United States could establish that the cost of business in the Arctic Ocean involves infrastructure support beyond traditional contingency planning. With additional government contribution, such precedent of public-private partnership could help to resolve many of the challenges of oil-spill response in the Arctic that were noted by the National Commission on the BP Deepwater Horizon Oil Spill and Offshore Drilling. More broadly, creating integrated response capacity for oil spills, safety of life at sea and other emergencies would be an efficient application of financial, political, and social capital.

The Bering Strait region south of the Chukchi Sea is home to indigenous communities in both Alaska and Chukotka. As the choke point into and out of the Arctic Ocean, at only around 50 miles across at its narrowest, the Bering Strait is a gateway region that requires emergency-response infrastructure with increasing urgency as commercial activities accelerate. As the closest connection between the United States and Russia, the region also offers a template for cooperative and consistent coordination among neighbors, which will further promote stability throughout the High North.

How can we go about doing this? Certainly, Congress could initiate an international, interagency, and public-private partnership to create emergency-response capacity in the Chukchi Sea. The United States would take the lead, but much like the International Space Station, this could be an enterprise that includes other international partners. Several U.S. agencies would want to participate, especially those involved with the Interagency Arctic Research Policy Committee and the Arctic Executive Steering Committee. Both the United States and specifically Alaska could partner with exploration-oriented companies in offshore leasehold areas to create the emergency-response platform. Admittedly, the major oil companies are not flush with cash, in light of lower oil prices, and Shell is slowing its exploration in the Chukchi Sea after disappointing tests. However, over the long term, such a platform in the Chukchi Sea, constructed as an international, interagency, and public-private partnership, could be a win-win for all. We should explore these possibilities over the coming months during the U.S. chairmanship of the Arctic Council.

12 "THE NAVY AND THE HIGH NORTH"

LCDR Rachel Gosnell, USN

In this brief but significant essay, Lieutenant Commander Rachel Gosnell reminds readers that it has been more than sixty years since 1958, when the USS *Nautilus* (SSN-571) conducted operations under the North Pole, and that the Navy has been present ever since in one manner or another. She also reminds readers that the Arctic is not a region that is easily reduced to a single entity. Rather, it is diverse and complex with respect to national interests, environment, population, economic activity, and weather patterns. The strategic interests of Russia and the United States are vastly different in the region, and the non-Arctic nations of China, South Korea, Japan, and the European Union have recently increased their activities and interests there, as have all eight Arctic Council nations. Balancing U.S. strategic interests in the region with constraints on naval resources is a significant challenge for which Gosnell offers eight succinct recommendations.

"THE NAVY AND THE HIGH NORTH"

By LCDR Rachel Gosnell, USN, U.S. Naval Institute *Proceedings Today* (August 2018).

The Navy has been present in the Arctic since the USS *Nautilus* (SSN-571) first ventured to the North Pole in 1958, albeit primarily in the undersea and air domains. While the region was of tremendous strategic importance during the Cold War, the years that followed saw diminished geopolitical interest. However, regional warming trends, combined with the potential for vast economic resources, have sparked global interest. While the Arctic has long been considered a peaceful and stable region, the Navy must seek to better understand the Arctic and to build greater capacity for operations there to ensure continued protection of U.S. strategic interests—and those of our allies—in the High North.

It is critical to examine strategic interests in the Arctic and develop a balanced approach to protecting U.S. interests globally. There is concern, particularly noted by Congress, that the United States is lagging in the Arctic. Yet it is critical to craft a thoughtful approach to the region that considers all U.S. strategic interests and naval resource constraints. To determine the most appropriate U.S. Navy strategy and policies for the region, it is necessary to understand the U.S. geostrategic, geopolitical, and economic interests, while also considering Arctic trends.

The Arctic is often simplified into a single region rather than fully appreciating the complexity and diversity it holds in terms of strategic interests, environment, population, economic activity, and weather patterns. There is no consensus on how to define the Arctic, though the most commonly accepted definition is the area north of the Arctic Circle (66 degrees, 34 minutes North). The Arctic Ocean is vast, encompassing approximately 5.4 million square miles, but it is the shallowest of the five major oceans. This renders much of the seabed accessible for exploration—and also impacts availability of deep-water ports and limits ship draft for the Northern Sea Route and Northwest Passage.

Since planting its flag on the bottom of the North Pole in 2007, Russia has ambitiously pursued its Arctic interests. There is no question that Russia holds significant interest in the region—nearly half of the Arctic's four million inhabitants and about half of the Arctic coastline are Russian. Increasingly reliant on the economic capacity of the Arctic, Russian derives 10 percent of its GDP and 20 percent of its exports from the Arctic zone. Strategically, an opening Arctic presents the opportunity for Russian maritime traffic to access the Atlantic and Pacific Oceans.

Russia has taken an assertive military posture to maintain its strategic and economic interests in the Arctic. The country recently held its largest alarm-exercise in a decade in the Arctic, consisting of 36 naval vessels and numerous aircraft and coastal weapons. The Northern Fleet—Russia's largest—consists of 41 submarines and 38 ships. In 2014, a new Arctic Command was established, consisting of the Northern Fleet, four Arctic brigade combat teams, 14 new airfields, and 16 deep-water ports. There are also significant infrastructure upgrades being made throughout the region. The 2014 Russian National Security Strategy only mentioned the Arctic three times, but the 2015 Maritime Doctrine of the Russian Federation devotes an entire chapter, "The Arctic Regional Priority Area," to it.

China is increasingly interested in the Arctic as well, releasing an Arctic Policy in January that formalizes China's intent as a "near Arctic state" to be active in the region with scientific research, natural resource exploration and development, and maritime trade. China is constructing its second icebreaker and has invested heavily in both Russia's Yamal LNG project and the new class of icebreaking LNG carriers designed to carry LNG from Sabetta to Asian markets.

South Korea, Japan, and even India have started to look northward to the vast economic resources in the region. European Arctic stakeholders have long been present, extending beyond the Arctic states to include France, Germany, and the United Kingdom. The European Union has also taken a proactive approach, releasing its Arctic Policy and appointing an Ambassador to the Arctic.

The U.S. approach to the Arctic has thus far been more muted. Although Secretary of Defense Mattis stated "the Arctic is key strategic terrain" in his confirmation hearing, there is no mention of the Arctic in the unclassified

National Defense Strategy and it appears only once in the National Security Strategy. It must be noted that U.S. strategic interests are vastly different, given the abundance of natural resources outside the Arctic (particularly with shale) that are easier and less costly to access. Shell's foray into Alaskan Arctic drilling proved disastrous, costing more than $7 billion before the company abandoned efforts because of the hostile environment and immense costs.

It is clear that U.S. strategic and economic interests in the Arctic differ greatly from Russia's, but the United States must be prepared to operate in all domains in the Arctic as activity increases. Though maritime activity is limited and will never rival sea lanes across the Atlantic or Pacific, traffic will rise as ice diminishes. Currently the Coast Guard leads national efforts in the maritime domain for the U.S. Arctic. The Navy must strive to understand and enable Arctic operations, while not shifting focus from greater threat regions.

Recommendations

There are a number of steps the Navy can take now and in the near future to develop Arctic proficiencies, without detracting from other critical missions. Doing so would be prudent for developing operational capacity in the Arctic as it becomes increasingly accessible. While activity in the undersea and air domains will undoubtedly continue, Secretary of the Navy Richard Spencer's call for blue water operations must be considered.

The Arctic is a unique operating environment, filled with immense challenges ranging from extreme cold—impacting personnel and machinery alike—to unpredictable weather, frequent fog, storms, and ice. Warming trends may diminish ice coverage, but it will remain challenging to operate in the region, and mariners must be skilled in the nuances of the Arctic. While it is unlikely the United States will operate large numbers of ships in the region in the near future, it must understand the requirements to do so and consider the following:

Improve Arctic communications and safety infrastructure. In a region notorious for poor communications and inadequate infrastructure, it is important to invest now in hydrographic surveys, navigation aids, and communications capabilities to ensure future operations are possible.

Prioritize personnel exchanges with the U.S. Coast Guard and partner navies operating in the High North. The Coast Guard has extensive experience operating in the Arctic and retains a cadre of officers who have honed this unique skillset. Embarking naval officers and midshipmen on Coast Guard vessels operating in the region would permit an enhanced understanding of regional challenges. Similarly, partners such as Canada, Norway, Sweden, and Finland have long engaged in Arctic maritime operations. Exchanging personnel could strengthen relationships with allies while also exposing U.S. Navy officers to vital Arctic skillsets.

Increase participation in High North exercises. The United States should seek to participate in bilateral and multilateral exercises, particularly with NATO allies and partners. There are a number of exercises the U.S. Navy already participates in, but additional assets could be sent to broaden exposure and training opportunities as well as improve capabilities with allies and partners.

Train to Polar Code standards. The development of the Polar Code by the International Maritime Organization reflects multilateral input to ensure safety of mariners. It recognizes the unique operating characteristics needed for the polar waters and requires specific training, qualifications, and cold weather precautions for vessels. Training to these standards would ensure Navy vessels could operate safely in the Arctic if called to do so.

Consider developing a framework for an Arctic Code for Unplanned Encounters at Sea (CUES). Given the success of CUES, implemented at the 2014 Western Pacific Naval Symposium, it could be helpful to extend CUES to the Arctic or develop a similar framework for Arctic states. While Russia, Canada, and the United States were party to this agreement, the other five Arctic states were not. This could help establish the Arctic maritime security environment as a regulated one, reducing the chance for a misperception or misunderstanding.

Improve Maritime Domain Awareness cooperation across the Arctic. As maritime traffic rises so will the occurrence of illicit activities such as illegal fishing, smuggling, narco-trafficking, and trans-criminal organizations. Much of this falls within the Coast Guard's law enforcement and constabulary responsibilities, but the Navy can also contribute to MDA and understand the Arctic maritime picture.

Enhance Security Cooperation in the Arctic through existing mechanisms such as the Arctic Security Forces Roundtable (ASFR). While Russia has been disinvited from the ASFR, the other Arctic states plus France, Germany, the Netherlands, and the United Kingdom are active participants. The Navy should further enhance participation in exercises of the Arctic Coast Guard Forum, developing its capacity for search and rescue should it be called north to respond to a crisis.

Enable Appropriate Presence in the Arctic region to protect U.S. interests, reassure partners and allies, and provide a strong deterrent. Presence primarily should be in the form of continued undersea and air operations, but can also include carefully selected training and exercises designed to improve regional familiarity, support allies, and increase deterrence.

Develop industry relationships with leading Arctic shipbuilders, particularly Scandinavian commercial and military industries, to inform future operational designs. While there are some calls for a purpose-built U.S. Arctic patrol craft, it would be imprudent for the Navy to have ships designed to operate primarily in the Arctic because of the global nature of U.S. sea power. However, the Navy's future frigate (FFG[X]) program should consider Arctic-specific operating requirements and incorporate these when practical—as long as they do not detract from the ability to operate in temperate waters where these vessels will be most needed.

While there is not a pressing need for U.S. Navy presence in the Arctic at present, the future will see a rise of maritime activity in the region. With other Arctic stakeholders developing their operating capacities for the region, the United States risks disadvantaging strategic interests if an appropriate foundation for future operations is not considered today. The U.S. Navy must act now to build the appropriate skillsets, operating abilities, and partnerships to enable future capabilities to operate in the High North to ensure the protection of national interests and maintain commitment to allies and partners.

13 "PROFESSIONAL NOTES: PREPARING FOR ARCTIC NAVAL OPERATIONS"

CDR Mika Raunu, Finnish Navy, and CDR Rory Berke, USN

In recent decades the U.S. Navy has sometimes operated in the Arctic, but it has not trained for Arctic operations or built ships for service in the Arctic. Mika Raunu, a Finnish naval officer, and Rory Berke, an American naval officer, provide an overview of the technical aspects of Arctic naval operations. In doing so, they highlight some of the unique challenges faced by naval surface operations that will affect winter doctrine and tactics at the task-force level and below.

"PROFESSIONAL NOTES: PREPARING FOR ARCTIC NAVAL OPERATIONS"

By CDR Mika Raunu, Finnish Navy, and CDR Rory Berke, USN, U.S. Naval Institute *Proceedings* (December 2018): 70–72.

On 19 October 2018, the USS *Harry S. Truman* (CVN-75) and ships in her strike group crossed into the Arctic Circle while operating in the Norwegian Sea in preparation for NATO Exercise Trident Juncture. It is the first time since 1991 a U.S. aircraft carrier has operated that far north, and Navy senior leaders indicated that in an era of renewed great power competition, operations in the high north will become more frequent and regular.

Indeed, U.S. and European navies certainly will be called on to operate in these waters more in the near future than they have in the recent past, as a warming Arctic sees more commercial maritime traffic and an increased Russian military presence. Then–Commandant of the U.S. Coast Guard Admiral Paul Zukunft said in August 2017 that the situation in the Arctic is similar to that of the contentious South China Sea.

The Arctic has not always been an area on the periphery for U.S. and NATO military operations. The United States relied on northern waters to deliver supplies during World War II. During the Cold War, the Arctic was a key area for strategic interests and power projection. And northern nations with enduring interests in the Arctic have built their navies to operate in this forbidding region, including incorporating civilian icebreakers to support naval operations.

In contrast, for more than two decades the United States has not built or trained its Navy to operate in the Arctic. Now as it returns there, it needs to get smart about the basics of operating in such a thoroughly unforgiving environment, especially in winter.

Basics of Winter Operations

Ships and weapon systems must be designed for winter conditions. Extremely cold weather places demands on ships and their equipment and, more important, on their crews. These crews must have a deep understanding of the constraints and characteristics of their ships and systems. The risk of breakdowns and other technical failures is higher in winter, and the consequences more serious. For example, if the ship's ice limitations are not observed, the risk of damaging the hull, rudder, or propulsion system while navigating in the icecap is higher. There is almost no margin for error in cold weather operations—if it can break, it will.

Winter conditions also affect the safety of those on board. The outer decks are vulnerable to icing, so clothing and footwear should be suitable for cold weather and slippery decks. Navies that do not often operate in extremely cold environments are likely to overlook the importance of properly equipping and clothing their sailors.

Meanwhile, extra effort should be made to ensure mission-essential equipment and systems will operate at full capacity in cold weather conditions. This means, for instance, that heating or deicing must be started and maintained while in a cold environment. And weapon systems should be protected by structural solutions that have been engineered during the design of the ship.

Effects on Warfare Areas

The winter operating environment in cold regions affects all warfare areas at least to some extent; it makes none easier and some much more complex.

Search and Rescue (SAR) Operations

SAR operations in cold climates require rapid actions because of the extremely short survivability times. In a man-overboard situation in the ice, sleds, ice awls, and immersion suits are crucial tools. Often the ship's helicopter is the best option for a recovery. Rigid-hulled inflatable boats can be difficult to maneuver in icy waters and often bear the brunt of winter weather because they are stowed exposed to the environment.

Air Operations

Cold weather affects the use of onboard helicopters and unmanned aerial vehicles (UAVs). The effects are greatest between 23 and 41 degrees Fahrenheit, when the dehumidification requirements from the systems are most demanding. Also, snowfall limits the use of helicopters and UAVs. Helicopters and UAVs should be kept in hangars with appropriate heating until they are needed. Ships without hangars are at a distinct disadvantage. Ship landing lights and other flight support devices and safety equipment must be kept clean of snow and ice.

Antisurface Warfare

Changes in visibility are significant in winter, making visual identification of surface contacts challenging. Snowfall and sleet hamper identification of traffic as well. Compacted ice can form stray echoes and limit the tracking of small targets. The effects normally are greater in littoral areas or in enclosed seas where the ice is easily compacted, such as in the Baltic Sea. Limited daylight hours must be considered when planning flight operations.

Antisurface missiles with radar seekers generally do not perform well in areas with sea ice. The ice and icebergs restrict their use and require very precise target tracking. However, antisurface missiles with imaging infrared seekers can be employed effectively.

The in-port missile loading process is far more demanding in winter. Ice blocks hamper the optimum berthing of the ship to missile loading piers and can make it difficult to mate the ship to the pier correctly. Once that is accomplished, even the smallest pieces of ice may prevent loading missiles into tubes and launchers.

Mine Countermeasures (MCM) and Mine Warfare

MCM ships are not icebreaking-capable because of their multiple hull-mounted sensors and antimagnetic materials. Some MCM ships can transit in ice with attached covers protecting the hull-mounted sensors.

Some MCM equipment, such as unmanned underwater vehicles, remotely operated vehicles, and clearance divers can be used under the ice. However, the risk of breaking equipment or system cables is high. MCM operations are an excellent example of why winter conditions must be considered during a vessel's design stage.

Arctic waters generally are shallow and thus suitable for sea mines. Ice does not restrict the use of influence mines in the seabed, but it strongly limits the use of moored contact mines. During mine laying, ice blocks can damage the horns or antennas of contact mines. After mine laying, drifting ice can move moored mines from their original positions and planned depths.

The hydroacoustic environment changes during winter and must be accounted for when programming acoustic settings on influence mines. Normally ice does not necessitate changes to pressure settings or magnetic sensors.

Winter conditions also hamper pier-side and shipboard mine loading. Forklifts and other logistics vehicles on board ships should be equipped with winter tires. Cranes for mine loading should be serviced with heating and lubricants designed for cold temperatures.

Antisubmarine Warfare (ASW)

It is difficult to use hull-mounted sonars or towed arrays in icy conditions, as both can easily be damaged. However, low water temperatures generally create optimum acoustics for ASW, even in shallow Arctic waters. Surface ships able to successfully deploy ASW sensors may thus be rewarded. The use of ASW torpedoes is possible, but ice limits torpedo launching. The risk of breaking propulsion systems, seekers, or wire-controlled optical fibers is high.

Because of the shallow waters, many countries use underwater arrays fixed in the Arctic seabed to monitor territorial waters or important areas. Navies that can convert contact detections to submarine prosecutions by cueing nearby surface ships have a significant advantage. Yet complex surface prosecutions of submarines remain extremely challenging in the Arctic. To maximize ASW, ships should use embarked or nearby land-based helicopters with dipping sonars and ASW torpedoes. It is possible to exploit non-ice-covered areas with ASW helicopters, but ice normally limits the use of antisubmarine airplanes.

Antiair Warfare

Winter conditions do not adversely affect shipboard air-surveillance radars or fire-control systems. However, depending on the locations of sensors relative to a ship's superstructure, deicing and heating may be required. For instance, radar antennas, especially those in the forward part of the superstructure, are vulnerable to icing. Fixed antennas with electronically scanned arrays can be particularly susceptible.

Snowfall and low visibility restrict target identification and the use of passive sensors such as TV cameras and infrared trackers. While proper preparation and equipment maintenance can mitigate this, long exposure to icy conditions will render such equipment minimally useful.

On most ships, antiair missiles are located near the forward part of the superstructure, where the missile systems and the ship's main gun are more vulnerable to icing. Snowfall restricts the use of missiles with infrared homing, but the limitations are smaller than those caused by rain. This should be taken into account when developing laser technology–based weapon systems for high latitudes.

Winter conditions also affect electronic warfare (EW) systems. EW sensor sensitivity changes with variations in temperature, as the scattering and reflection of radio waves in the atmosphere is more pronounced in a cold environment.

Navigation

Ships transiting ice fields normally must do so at slower speeds. The shortest route option usually is not the fastest. It is beneficial to exploit open water whenever practicable.

While Western navies have worked for years to integrate UAVs for targeting or intelligence, surveillance, and reconnaissance missions, in the Arctic UAVs and helicopters are more ideal for ice scouting in search of optimum routes.

Cooperation with maritime patrol and reconnaissance aircraft, other ships, and icebreakers is vital to maintain an accurate operational picture and determine if icebreaking assistance is required. Cooperation with coast guards and civilian ice services and icebreakers is important. In Finland, navy and coast guard ships routinely report the ice situation and observations of icing to civilian authorities, who forward the information to the broader maritime community. Similarly, civilian icebreakers regularly cooperate with regional navies.

Before departing port, crews must assess the situation along the route or in the planned operating area. Ice thickness must be monitored to ensure the ship's operating limits are not exceeded. Crews should take special equipment on board to assist with winter navigation, including snowmobiles, spare UAVs, traditional skis, and ice drills.

Engineering Branch

It is essential to adjust the ship's draft and trim so the ice-strengthened part of the hull is at an optimal level. The cooling-water pipelines and engines, air conditioning, and hydraulics all must be operational in winter temperatures.

Normally, stabilizer fins cannot tolerate collisions with ice blocks and must be either detached or positioned inside the hull to ice-transit positions.

Cranes and other deck gear must be protected from icing. If possible, their heating should be switched on. All lubricants must be designed for cold temperatures. In winter conditions where a ship is exposed for a long time, rubber and wooden sledgehammers are required to remove extra ice load. Cold temperatures also affect the nuclear, biological, and chemical (NBC) systems on board. The use of fixed NBC sprinklers can increase the ice load on decks.

Better Winter Navies

NATO's 2007 Naval Arctic Manual is an excellent guide. It includes extensive information in compact form. However, more specific winter doctrine, tactics, and manuals should be developed for the task-force level and lower. Too many lessons are shared unofficially and never written into publications.

As the demand for Arctic operations increases, cold-weather training must be increased. Navies with Arctic capabilities and experience should regularly exercise with others interested in building similar capabilities. In the same way the U.S. Ice Exercise aims to advance the operational readiness of submarines, surface forces should train together. While Western navies contemplate navigating in potentially contested Pacific waters, they should not neglect what it will take to regularly operate above the Arctic Circle, where ice, snow, and extreme cold present a unique set of challenges.

14 "TO THE NORTH POLE!"

Dr. Dean C. Allard

U.S. Navy operations in the Arctic date to 1849 and the search for Sir John Franklin, whose expedition disappeared in 1845. The American lead for this endeavor was Navy lieutenant Edwin De Haven, who was sent to the Arctic after Lady Jane Franklin made a personal appeal to President Zachary Taylor. Although De Haven had only modest results, his actions reignited American interest in the theory of an open polar sea that Matthew Fontaine Maury had championed. Dean C. Allard's article provides an excellent study of early U.S. naval operations in the region and demonstrates that present-day interest is the continuation of a long-standing desire for successful maritime and naval operations in the far north.

"TO THE NORTH POLE!"

By Dr. Dean C. Allard, U.S. Naval Institute *Proceedings*
(September 1987): 56–65.

The genesis of the U.S. Arctic programs was an 1849 appeal from Lady Jane Franklin to President Zachary Taylor for assistance in searching for a British naval expedition commanded by her husband, Sir John Franklin. The disappearance of this party—last seen by Europeans near Lancaster Sound in 1845—led to scores of rescue attempts over the coming decade. Most were mounted by the British. But Lady Jane's petition resulted in an American effort under the command of U.S. Navy Lieutenant Edwin Jesse De Haven.[1]

De Haven's ships—the *Advance* and *Rescue*—were provided by Henry Grinnell, a prominent American merchant and a student of world geography. In honor of his benefaction, this enterprise became known as the First Grinnell Expedition. Grinnell's ships, however, were Navy-manned and equipped at government expense.

In the summer of 1850, De Haven's party departed the United States, proceeded through Lancaster Sound, and joined British units searching for Sir John Franklin. After sharing in the discovery of the British explorer's first winter camp at Beechey Island, De Haven sailed north through Wellington Channel, one of the routes outlined in Franklin's original instructions. At this point, the *Advance* and *Rescue* became frozen in the ice and were driven to the north. They sighted a new territory, now known to be a westward extension of Devon Island, which De Haven named Grinnell Land. Later, the flow of the ice pack reversed, forcing the ships eastward to Baffin Bay. The expedition was finally freed from the grip of the frozen sea in the summer of 1851, but by that time unusually severe conditions in the Canadian archipelago were reported. That factor, plus concern over the health of his scurvy-ridden crew, caused De Haven to end his cruise.

While the tangible results of this expedition were modest, they were rich in significance for the future of U.S. polar activities. Even though his ships did not advance north of the 76th parallel, De Haven's experiences renewed interest in the theory of the open polar sea advanced by Matthew Fontaine Maury and other authorities. Maury, a famed U.S. Navy hydrographer, had special influence on American thinking. He based his hypothesis upon the correct observation that the coldest regions in the northern hemisphere are south of the Arctic Ocean. Maury also supported his belief in a large body of open water upon reports that whales and other air-breathing animals frequented the far north.

Regarding the search for Franklin, Maury's views led to the belief that the British explorer may have taken advantage of an unusually warm summer to enter the unfrozen polar sea but was prevented from returning to civilization by the girdle of ice in lower in latitudes. Searchers hoped that Franklin and his men were surviving by hunting and fishing. Maury's speculation provided a powerful impetus for future American expeditions to head for the North Pole in their quest for the Franklin party.[2]

Another long-range influence of De Haven's operation was the appearance of a notable new U.S. explorer, Elisha Kent Kane, who had served as a naval surgeon with the First Grinnell Expedition. Despite a serious heart ailment resulting from rheumatic fever, Kane was determined to live life to its fullest. After establishing himself as one of De Haven's most resourceful officers, the young surgeon returned to the United States and delivered a series of lectures that made him a well-known public figure. Of greater significance, Kane was also determined to return to the Arctic to seek the open polar sea, which he was convinced held the key to Franklin's disappearance. Incidental to this operation, he spoke of reaching the North Pole. He was the first American explorer to embark on this objective, which dominated the American imagination for many years.

Kane's expedition received both governmental and private support. Kane served under special orders from the Secretary of the Navy, which allowed him full pay. The Navy provided other officers and men, although they were supplemented by civilian scientists and Hans Hendrik, a young Eskimo hunter from Greenland. Henry Grinnell once again offered the *Advance*. Grinnell also obtained funds to equip the cruise. The enterprise became known as the Second Grinnell Expedition.

The northward route that Elisha Kane pioneered, later known as the "American route to the Pole," started in Smith Sound at the northern end of Baffin Bay. In the early 1850s, this area marked the northern limit of European geographic knowledge of this region. Kane planned to establish a base above Smith Sound from which dog-driven sledge and boat parties could reach the open waters that presumably lay beyond the masses of ice extending to approxi-mately the 80th parallel. Although Kane's plans were based on Maury's teachings, he was also indebted to recent observations by the British explorer,

Edward A. Inglefield. In 1852, while leading another Franklin rescue party, Inglefield became the first Western mariner to sail through Smith Sound. At that time, Inglefield reached Cape Sabine on Ellesmere Island at 78° 28′ north latitude and observed a large ice-free bay to the north. This sighting suggested the exciting possibility of an easy route to the pole.

Kane's achievements included driving the *Advance* through heavy ice to a new far-north position on the coast of Greenland at 78° 37′ north latitude. At this point, which Kane named Rensselaer Harbor, the ship was frozen into the ice. But, as planned, the expedition then journeyed by sledge to establish supply caches for use the following spring when a party with small boats would be sent to the open Arctic Ocean. During these operations, Kane became the first white man to encounter the Eskimos of the Smith Sound region, including those from the villages of Etah and Anoatok, who were destined to play a critical role in future polar expeditions in this area.

Kane could not achieve his most ambitious objectives because most of his dogs died from disease and bears plundered his food depots. Nevertheless, he and his men charted the perimeter of a large, new area, which came to be known as the Kane Basin, and were the first to sight the enormous Humboldt Glacier. A sortie made by William Morton, one of Kane's seamen, and the young Eskimo Hans Hendrick reached Cape Constitution on the Greenland coast at approximately 80° 58′ north latitude in June 1854. At the northern edge of the Kane Basin, Morton and Hendrick sighted another narrow, ice-free passage leading toward the pole. That body of water, believed to be an arm of the open Arctic sea, was named Kennedy Channel in honor of the U.S. Secretary of the Navy John Pendleton Kennedy.

During the next two winters, Kane's ship remained frozen in the ice off Greenland. The crew insulated the *Advance* with moss, grass, and snow, which offered some protection from the extreme cold. But in 1855 the expedition abandoned hope of freeing the brig. At this point, Kane left his base at Rensselaer Harbor and sought safety in the Danish settlements of southern Greenland. This trek by sledge and small boat, which depended upon the inspired leadership of Kane with assistance from the Smith Sound Eskimos, did much to guarantee the expedition's fame.[3]

One of the Second Grinnell Expedition's notable legacies was Kane's development and application of Arctic travel and survival techniques. Learning from the experience of British Arctic parties and his own observation of Eskimo methods, Kane became an American pioneer in using dogs, sledges, and pre-established caches for movement through the polar region. He was also one of the first explorers to heed the Eskimos' preference for undertaking long journeys during the winter, when the ice and snow were firmest. Kane's study of Hans Hendrik's hunting practices convinced him that it was possible to live off the land in the Far North. As a medical doctor who faced the problem of scurvy during the First and Second Grinnell Expeditions, Kane further recognized that the native diet—notably the consumption of raw meat—offered an effective preventive to that disease. Finally, Kane adopted Eskimo construction practices in protecting his ship from the rigors of the Arctic winter.

Kane's navigational practices were of additional interest for students of later polar expeditions. The positions determined by fixed equipment near Kane's base in Rensselaer Harbor were highly accurate. Yet, when his advance parties left that area, the limitations of their portable sextants, artificial horizons, and pocket chronometers became evident. For example, because of the sun's relatively flat, low path at high latitudes, it was difficult to determine precisely the apogee of the sun at local noon. In triangulating the Kane Basin, Kane and his men also depended upon an erroneous base line and relied to some extent on dead reckoning, which was "hardly more than a guess" in the "rocky shores and the rough ice belt" of the Kane Basin. It is not surprising that the expedition's reported positions in field were in error by approximately 30 to 50 miles.[4]

Kane died of heart failure at the age of 37, but his fame propelled explorers to continue his polar quest. One of these was another medical doctor, Isaac I. Hayes, who served his Arctic apprenticeship under Kane in 1853–55. As Hayes began to plan his own operation in the mid-50s, evidence finally came to light indicating that John Franklin and his men had perished in the Canadian Arctic during 1847–48. With the humanitarian objectives gone, Hayes stressed the need to continue Kane's effort to reach the open polar sea and to collect scientific data on the region north of Smith Sound.

After five frustrating years, by 1860, Hayes was able to obtain the financial support of a group of private benefactors. Not surprisingly, one of these was Henry Grinnell. Purchasing a small schooner that he named the *United States*, Hayes departed from Boston in July 1860. He called on the Greenland settlements and added Eskimo dogs and hunters to his party, including Hans Hendrik. As Hayes continued to sail along the coast of Greenland, he encountered heavy ice conditions and was frozen in for the winter off Etah, some 80 miles southwest of Kane's Rensselaer Harbor. Nevertheless, from this location, Hayes investigated the periphery of the Greenland ice cap and then crossed Smith Sound to reconnoiter the Ellesmere coast. He reached the approximate position of 80° 11′ north latitude, but, to his disappointment, did not sight the open polar sea.[5]

A more famous polar explorer of this era was Charles Francis Hall. This colorful engraver and newspaper publisher became fascinated with the north after reading of Elisha Kane's exploits. Hall also continued to be intrigued by Sir John Franklin and his party, despite the apparently reliable reports of their demise in the region of King William Island. Hall's first two visits to the north were primarily to investigate the fate of Franklin's expedition.

For the fourth time, Henry Grinnell came to the financial aid of northern exploration by raising funds for Charles Hall. With these resources, Hall booked passage on a whaling ship that departed from New London, Connecticut, in 1860 and proceeded to Baffin Island. During the next two winters, as the ship became frozen in the ice, Hall lived much of the time with the Eskimos of the region, developing a deep friendship for them and an admiration for their ability to survive in a seemingly hostile land. However, Hall was unsuccessful in persuading the Eskimos to travel with him to King William Island before he returned to the United States in 1862.

From 1864 to 1869, Hall once again journeyed to the Arctic. This time, his base was the Melville Peninsula north of Hudson Bay. He lived with the Eskimos in that region for most of this period and became proficient in native hunting and travel techniques. In 1869, Hall succeeded in reaching King William Island by sled. He discovered relics belonging to Franklin's men, before they perished from starvation and exposure more than 20 years earlier, and confirmed the unlikelihood of survivors living in the area.

These lengthy visits to the far north prepared Hall for his great life's work—a voyage toward the North Pole. Hall persuaded Congress to authorize an official expedition administered by the U.S. Navy and supported in its scientific aspects by the Smithsonian Institution. The expedition's vessel was a 140-foot naval tug, the *Polaris,* specially modified for Arctic service. This sturdy ship departed from the United States in July 1871 with a crew of 25 officers, men, and scientists. In addition, eight Eskimos, including the experienced Hans Hendrik, joined the ship to act as hunters and sled drivers.

Although troubled by cases of increasing insubordination by several members of the party, the *Polaris* was fortunate to encounter an unusually mild Arctic summer. Hall navigated with little difficulty through Kane Basin and Kennedy Channel and finally through two bodies of water never before seen by nonnative travelers. They named these Hall Basin and Robeson Channel, in honor of the expedition's commander and the Secretary of the Navy, George Maxwell Robeson. As the *Polaris* reached the northern end of Robeson Channel at approximately 82° 11′ north latitude, she faced the Arctic Sea. But any belief that this would be an open expanse of water soon disappeared. Instead, Hall and his men viewed a frozen surface stretching as far as the eye could see.

At this point, weather conditions deteriorated, and heavy winds and masses of ice drove the ship to the south. When it appeared that the *Polaris* might be crushed, Hall was able to reach relative safety in a small bay on the Greenland coast at 81° 37′ north latitude. With proper gratitude, he named this position Thank God Harbor. They covered the ship with snow and ice to protect the crew for the winter season, much as Elisha Kane had enclosed the *Advance.* Then, in October 1871, Hall and three companions went on a two-week reconnaissance by sled to the north to prepare for a polar attempt the following spring. After moving along the Greenland coast, the party climbed a high point of land, from which they could see the northern tip of Ellesmere and the eastward tendency of the Greenland shoreline. This led Hall to speculate that both positions were islands, rather than southward extensions of a major polar landmass.

If Hall had lived, he might have been the first to confirm the insularity of Greenland and Ellesmere and to undertake a long journey on the polar sea ice. But, immediately after returning to his ship, Hall was stricken with an illness. He died two weeks later. An official U.S. Navy investigation eventually concluded that Hall had suffered a stroke, even though the explorer had suggested in his final days that he had been poisoned by a malcontent crew member. Amazingly, this view was tested scientifically in 1968, 97 years later, when Hall's frozen body was examined in its icy grave at Thank God Harbor by a party directed by Professor Chauncey C. Loomis. Body samples taken during this long-delayed *post mortem* were evaluated in a Canadian laboratory and revealed that Hall had, indeed, ingested fatal amounts of arsenic during the last weeks of his life. Loomis noted that this poison may have been self-inflicted since arsenic compounds were common forms of medicine in the 19th century and Hall had rejected the care of the expedition's surgeon, Emil Bessels, with whom he had a strained relationship. Nevertheless, if Hall was murdered, Dr. Bessels must be considered the prime suspect.[6]

In tracing the continuing polar quest, we must turn to an effort to reach far northern latitudes via the Bering Straits. This route had special fascination in the 1870s because explorers believed that the warm Japanese current, penetrating deep into the Arctic basin, opened an easy sea route to the pole. In seeming contradiction to this new version of an open polar sea, the German geographer, August Petermann, also postulated the existence of a major continent north of the Siberian coast. Petermann argued that the landmass stretched across the Arctic basin, presenting a land bridge that could lead explorers to high latitudes.[7]

The leader who set out to test these theories was another American naval officer, Lieutenant Commander George Washington De Long, whose polar interests had been aroused when he led a small-boat voyage to northern Greenland waters during 1873 in search of Charles Hall's party. De Long was the last American explorer to receive advice and assistance from Henry Grinnell. Shortly before Grinnell's death in 1874, he suggested that De Long seek the support of the American newspaper magnate, James Gordon Bennett. As the publisher of the popular *New York Herald,* this wealthy young man was always seeking interesting copy, and he agreed to endorse De Long's efforts. By the late 1870s, Bennett and De Long acquired a steam bark, soon renamed the *Jeanette*, and obtained congressional approval for an expedition from the Pacific toward the North Pole. Although the expedition was administered by the Navy, Bennett bore most of the expenses.

In the summer of 1879, the *Jeanette* sailed from San Francisco and shaped a course for Wrangel Island. Explorers had suspected this territory existed long before Western mariners sighted it in 1867 when an American whaling ship succeeded in reaching the area through the heavy ice normally found off the Siberian coast. Petermann and other authorities were convinced that Wrangel was part of the presumed polar continent.[8] Therefore, De Long planned to establish a base in that area from which to mount an overland sledge journey to the top of the world.

The *Jeanette* was soon locked in the ice near Herald Island, but De Long expected his ship to drift to Wrangel, which was only 40 miles eastward. To his dismay, the speed of the polar pack made landing an impossibility. And, as the ice propelled the *Jeanette* past Wrangel, the crew realized that this position was a relatively isolated island. In the succeeding 21 months, the ship continued a long drift through the Arctic Ocean. Observations during this period revealed no other large landmass in the shallow seas off Siberia, although they discovered three small islands. Finally, in July 1881, at a point north of the 77th parallel, the *Jeanette* sank after being crushed by the polar ice, and her crew was forced to make a hazardous retreat by sled and small boat to the Siberian mainland. During this phase of the operation, De Long and a number of his compatriots died. Despite its tragic ending, however, the expedition helped disprove facile theories of an open polar sea and of the presumed Arctic landmass near Russia, and provided valuable insight into the polar basin's hydrography. Some years later, the discovery of wreckage from the *Jeanette* near Greenland offered further evidence of the prevailing currents in the Arctic Ocean.

No American attempted to reach the pole during the 80s. The 1890s were dominated by the obsessive determination of Robert Edwin Peary to become the conqueror of the pole. Peary was a U.S. Navy civil engineering officer, although he was on leave of absence during most of his northern expeditions. His famous dictum that "the more dramatic your expeditions are, the more incompetent you are," indicated the high degree of planning and order that he introduced to polar exploration.[9] His methods became known as the "Peary System."

Peary's remarkable persistence is shown by his repeated visits to the Arctic for more than 20 years.[10] On his first trip in 1886, he made a relatively short journey on the Iceland ice cap. Beginning in 1891, he concentrated on exploring northern Greenland in anticipation of using that area as a base for a polar assault. During this period, Peary mistakenly identified Peary Land as a separate position, rather than as a continuation of Greenland. Four years later, the explorer made his first attempt to cross the tortured surface of the frozen Arctic Ocean, followed by a similar expedition in 1899. But, in 1902, Peary's now considerable experience with the relatively rapid eastward drift of the ice off Greenland caused him to shift his starting point for the pole to Ellesmere Island. During that year, he reached 84° 17′ north latitude before he was forced to return to the coast. Throughout all of these operations, Peary depended upon the assistance of the Smith Sound Eskimos, who were integral members of his exploring teams.

Following the rebuff of 1902, Commander Peary returned to the United States to plan further operations. With financing provided by the Peary Arctic Club, composed of a number of wealthy Americans who continued the tradition of support from the American business community exemplified by Henry Grinnell, Peary built a superb steam schooner, the *Roosevelt*. She was specially designed for Arctic operations. In 1905, the *Roosevelt* carried Peary's party to Cape Sheridan on the northern coast of Ellesmere. Early the next year, Peary set out once again across the polar ice, only to be stopped by a large lead of open water at 87° 06′ north latitude. On his return to the Canadian archipelago, Peary made another erroneous observation, when he claimed to sight a territory, now known to be nonexistent, that he named Crocker Land.

In 1908, at the age of 52, Peary began his final attempt on the North Pole. After forcing the *Roosevelt* through the summer ice to Cape Sheridan, overland parties established a base at Cape Columbia, some 100 miles to the north-west. This position at 83° 07′ north latitude marked the northernmost point on the American continent and was only 413 miles from the pole. From his large party, which included 50 Eskimos and approximately 250 dogs, Peary formed five separate sledge teams; he headed one himself. The overall plan called for these groups to move in relays, starting in March 1909, to break a path and lay down caches of food and fuel. Slightly below the 88th parallel, on 1 April 1909, the last of the advance parties turned back. At that point, accompanied by the strongest Eskimos and dogs and Matthew Henson, the most capable non-Eskimo member of his expedition, Peary set out to achieve his lifelong goal. His report showed that he traveled the remaining 133 miles to the pole, reaching it on 6 April. He then made a rapid return to Ellesmere in order to reach the coast before the spring thaws began to break up the polar ice.

The Peary System's success resulted from a number of factors, not the least of which was the expert assistance provided by the Eskimo sledge drivers. Although Peary did not depend primarily upon hunting for his rations and lacked the cultural sensitivity for native peoples that was typical of many other Arctic explorers, he willingly adopted many Eskimo techniques. For example, in order to minimize the weight carried by his teams, heavy tents and sleeping bags were replaced by Eskimo-built igloos and special clothing made by Eskimo women. A few simple modifications allowed these clothes to become sleeping gear at night. Peary also refused to carry small boats; he was convinced that during the cold season chosen for his travels any leads encountered would soon freeze over. Each of his men, however, carried a sealskin bag. As the Eskimos had known for many years, these could be inflated and used as emergency floats to ferry sleds across open water.[11]

Peary's navigational techniques did not differ materially from those of his 19th century forebears. He relied primarily upon latitude observations based on measuring the sun's extreme altitude with a hand-held sextant and artificial mercury horizon. By timing the sun's daily apogee or determining the sun's true bearing at this point, he could also determine longitude, but he did not appear to make these computations while on the polar ice. In addition, Peary used odometers and compasses for dead reckoning. This imprecise technique, however, was not the primary basis for Peary's claim to reach the pole.[12]

The accuracy of Peary's navigation is open to serious question, particularly because he used hand-held equipment. Peary was also faulted for failing to take longitudinal observations. Under these circumstances, even those authorities who support Peary as the substantive discoverer of the pole are unwilling to guarantee that he reached that precise position.[13]

Shortly before Peary announced his achievement, a former associate, Dr. Frederick A. Cook, stated that he had reached the top of the world one year earlier—in April 1908. Cook's account revealed an expedition of great daring and extent. Lacking his own ship or the other material and human resources of Peary's large party, Cook made a long, difficult sledge journey from the Smith Sound area to Axel Heiberg Island, with a small group of companions. His report indicates that he departed Axel Heiberg in March 1908 and traveled some 520 miles across the frozen sea's rugged surface, accompanied by two young Eskimos. After reaching his goal, Cook returned to the Canadian archipelago and spent the winter of 1908–09 on Devon Island. During his expedition, Cook sighted a territory that he named Bradley Land which, like Peary's Crocker Land, is now known to be mythical. Throughout much of this 14-month operation, Cook lived off the island.

For navigational guidance, Cook attempted to determine longitude by timing the sun at its extreme altitude, hence the results of these computations. He also made limited use of shadow observations and depended extensively upon dead reckoning techniques. Nevertheless, as was true for Peary, his proof of reaching the pole depended primarily upon latitude computations derived from the somewhat imprecise observation of the sun's altitude with hand-held sextants and mercury horizons.[14]

It is extremely difficult to prove scientifically the reports of either explorer. Neither was accompanied by other navigators who could take confirming observations. Nor could later expeditions validate the physical evidence Peary and Cook left, since their cairns had long since drifted away on the polar ice. In the last analysis, therefore, proponents of Peary and Cook must depend on the credibility of these leaders, reminding one of Norwegian Roald Amundsen's astute observation that the "character of the explorer . . . is always the best evidence of his claim of achievement."[15] Character judgments, however, are highly subjective and are unlikely to command universal acceptance.[16]

Although the Peary and Cook expeditions are of major interest, other American polar explorers during this period preferred Spitsbergen and Franz Josef Land as bases of operations. One of these individuals was a colorful journalist, Walter Wellman, who in 1894 and 1899 made short passages on the sea ice from both of these islands. In 1907 and 1909, Wellman, apparently inspired by Salomon A. Tree's pioneering balloon attempts in the late 1890s, returned to Spitsbergen with a large airship named *America*, which made two unsuccessful flights.

During the succeeding decade, the world's major nations were absorbed by World War I, and most Arctic exploration was put on hold. In the 1920s, when interest in reaching the pole resumed, aviation technology, which had improved dramatically since the early part of the century, received special attention. Roald Amundsen, the Norwegian discoverer of the South Pole, working with Lincoln Ellsworth, an American, demonstrated the feasibility of aerial operations in 1925 by flying from Spitsbergen in two fixed-wing aircraft to within approximately 120 miles of their goal. In the next year, another Amundsen-Ellsworth flight was made in the Italian-designed airship *Norge*. During a 72-hour flight, this expedition not only reached the top of the world, but crossed the entire Arctic basin and landed near Nome, Alaska.[17]

Only two days before Amundsen began his successful flight, his ambition to become the first aviation discoverer of the North Pole was preempted by Richard E. Byrd and his assistant, Floyd Bennett. Flying from King's Bay, Spitsbergen, in a three-engine Fokker aircraft, these pioneers reported reaching 90° north latitude on 9 May 1926. As was true for so many other American explorers, Lieutenant Commander Byrd was a naval officer. During his discovery flight, however, he was on leave from the service, and his trip was financed by private sources.

Considering the controversy over the positional accuracy of previous Arctic expeditions, it is not surprising that Byrd gave special attention in his reports to navigational details. He noted, for example, that the sun compass, a device projecting the sun's shadow onto the hands of a 24-hour clock, determined the aircraft's longitude and allowed them to steer a due northern course. The speed of advance along this meridian was determined by dead reckoning. When they reached the pole, they obtained confirmed sun sights by sextant and artificial horizon. Byrd's precise return to his takeoff point in Spitsbergen was another indication, in the aviator's estimation, of the accuracy of his dead reckoning.

Five years later, an underwater approach to the pole, discussed hypothetically since the 19th century, was attempted for the first time by Sir Hubert Wilkins. Although Wilkins was Australian, his submarine, the *Nautilus*, had been a U.S. Navy vessel. In addition, many individuals associated with the expedition were American nationals, including Simon Lake, the original designer of the *Nautilus* who modified the craft for Arctic duty; and Sloan Danenhower, a former U.S. naval officer whose father had served with Lieutenant De Long during the *Jeanette* operation. Danenhower commanded the submarine under Wilkins's general direction and insisted on maintaining positive buoyancy in order to allow the submarine to slide along the bottom of the ice on inverted runners and to surface automatically in leads and polynyas to replenish her air supply. A large screw installed on the superstructure was designed to bore an airway in case they encountered solid ice. Unfortunately, the inability to retract this screw at a critical point; the failure of the submarine's diving planes, possibly as a result of crew sabotage; and inherent limitations in the submarine's capabilities restricted the *Nautilus* to a short cruise under the polar pack near Spitsbergen. Nevertheless, as was true for the pioneering aeronautical attempts of Salomon Andree and Walter Wellman, the primitive *Nautilus* expedition later inspired more successful operations that could take advantage of advances in technology.

Following World War II, aerial navigation over the pole, including regularly scheduled commercial flights, became commonplace. Another activity during this period was the habitation of large ice islands as they drifted through the polar sea. By establishing scientific stations on these platforms, they became, in effect, the 20th century equivalents to De Long's *Jeanette*. Probably the most famous U.S. ice station was manned by an Air Force team under the direction of Colonel Joseph O. Fletcher. This island, known as T-3, was only 103 miles from the pole when Fletcher arrived in 1952, and he hoped the natural drift would lead it across the 90th parallel. Although this did not happen, they collected valuable oceanographic and meteorological observations over the next two years. Later, other ice islands manned by Air Force teams and by parties associated with the U.S. Navy's Arctic Research Laboratory at Point Barrow, Alaska, yielded additional scientific data that increased human knowledge of the polar regions.[18]

The quest for the North Pole also continued to dominate the imagination of U.S. explorers. In 1968, an unusual expedition claimed to be the first to reach the North Pole over the surface since the days of Cook and Peary. Ralph Plaisted, a businessman from St. Paul, Minnesota, led this party, which traveled by 12 horsepower snowmobiles from Ward Hunt Island off the Ellesmere coast. The party received supplies from supporting aircraft. Plaisted's team reached the objective on 19 April and later returned to its starting point by air. Air Force planes, using modern navigational equipment, verified Plaisted's achievement.[19] Eighteen years later, a team of six U.S. and Canadian adventurers, led by Will Steger of Minnesota, reached the North Pole from their base at Ward Hunt Island.[20] In this 56-day, 500-mile trek, they directly imitated Peary by relying solely on dog-driven sleds.

Another notable operation in the Arctic occurred in 1969 when two U.S. Coast Guard icebreakers, the *Staten Island* (WAGB-278) and *Northwind* (WAGB-282), and the Canadian icebreaker *John A. MacDonald* escorted the tanker *Manhattan* on a westward voyage through the Northwest Passage. The Coast Guard's deep knowledge of navigation in this region was also demonstrated by the role of its representatives in designing a special down-breaking ice bow for the 150,000-ton tanker and the invaluable aerial reconnaissance of the ice-choked passages of the Arctic Archipelago provided by Coast Guard aircraft operating from Greenland. This pioneering transit by a large merchant vessel through the legendary Northwest Passage demonstrated the feasibility of using these waters for commercial purposes, including the shipment of oil produced in the far Arctic regions.[21]

A dominant aspect of post–World War II polar exploration was the renewal of submarine operations. They had been largely abandoned after Wilkins's attempt. These voyages became possible because of the almost limitless underwater range of nuclear propulsion plants and dramatic improvements made in navigational equipment.

Appropriately, this story begins with the USS *Nautilus* (SSN-571), the world's first nuclear-powered submersible.[22] In 1957, that vessel, commanded by William R. Anderson, made a polar attempt north of Spitsbergen.

Damage to a periscope, suffered while surfacing in the polar pack, and the failure of the vessel's gyrocompass, forced the *Nautilus* to turn back at the 87th parallel. But, in April and July 1958, the submarine returned to the Arctic, this time using the Bering Straits approach. The April operation was blocked by the shoal water of the Chukchi Sea and ice ridges that extended to unexpected depths. In July, however, the *Nautilus* found a submerged sea valley in the Cape Barrow area that presented a deep-water passage beneath the polar pack. Then, in a classic 96-hour, 1,800-mile underwater voyage across the largely uncharted Arctic Ocean, the submarine reached the pole and continued on to the edge of the polar ice north of Greenland. This cruise demonstrated the feasibility of a new merged sea route that reduced the distance from London to Tokyo by 4,700 miles. Commander Anderson also gave credit to the instrumentation that was essential for his operation's success. Of particular importance was sonar equipment used to trace the profile of the overhead ice. In addition, inertial navigational instrumentation, improved gyrocompasses, and the more traditional techniques of dead reckoning were of major importance.

During the summer of 1958, another nuclear submarine, the USS *Skate* (SSN-578), commanded by Commander James Calvert, reached the pole from the Spitsbergen area.[23] In addition, this vessel surfaced in the leads and polynyas that are common in the polar pack during the summer. In March 1959, the *Skate* returned to the region to determine if she could also ascend through the frozen sea in the coldest month of the Arctic season. By choosing areas of thin ice known as greenhouses from their underwater appearance, using closed circuit television to observe the overhanging pack, and by employing her sail area as a battering ram, the *Skate* was able to surface a number of times. Navigationally, these ascents were significant since they allowed the submarine to take sun and star sights to confirm locations obtained from dead reckoning and inertial navigational equipment.

In February 1960, the USS *Sargo* (SSN-583), commanded by Lieutenant Commander J. H. Nicholson, used the Pacific approaches to become the third submarine to reach the pole. Then, in August and September of the same year, the USS *Seadragon* (SSN-584), commanded by Commander George P. Steele, made another Arctic cruise, which originated in the Atlantic. This cruise included the first submerged transit of the Northwest Passage before the vessel proceeded to the pole. She also demonstrated her ability to operate safely in the area of icebergs, despite the notorious instability and enormous depths of these formations.[24]

Looking back at the many decades of U.S. polar activity, an observer must be impressed by the continuity of these efforts. To be sure, the humanitarian concerns of early expeditions seeking to aid Sir John Franklin were set aside when the tragic fate of his party became known. Yet, from the earliest days, American explorers also had a genuine interest in geographic and scientific discovery, and this concern is equally evident today. The use of Eskimo techniques of Arctic survival and travel, a prominent aspect of the early period of exploration, was superseded by the application of advanced technology in the 20th century. Whatever methods are used, the need for special adaptation to the exotic Arctic environment is another continuing strand in the quest for the North Pole.

Notes

1. De Haven's expedition is discussed in John Edwards Caswell, *Arctic Frontiers: United States Explorations of the Far North* (Norman, OK: University of Oklahoma Press, 1956), pp. 13–20 and in George W. Comer, *Doctor Kane and the Arctic Seas* (Philadelphia, PA: Temple University Press, 1972), pp. 71–101.

2. Comer, *Doctor Kane and the Arctic Seas*, pp. 79–80; Frances L. Williams, Matthew Fontaine Maury, *Scientist of the Sea* (New Brunswick, NJ: Rutgers University Press, 1963), p. 202.

3. Farley Mowat, *The Polar Passion* (Boston, MA: Little Brown, 1967), pp. 103–104, stresses Eskimo contribution during the Kane expedition. Oscar M. Villarejo, *Dr. Kane's Voyage to the Polar Lands* (Philadelphia, PA: University of Pennsylvania Press, 1965), pp. 53–55 and passim comments on dissension within Kane's party.

4. Corner, *Dr. Kane and the Arctic Seas*, pp. 259–63.

5. Caswell, *Arctic Frontiers*, pp. 32–41. See also Hayes's own account, *The Open Polar Sea* (New York: Hurd and Houghton, 1869).

6. Chauncey C. Loomis, *Weird and Tragic Shores: The Story of Charles Francis Hall, Explorer* (New York: Alfred A. Knopf, 1971) pp. 336–54.

7. Caswell, *Arctic Frontiers*, pp. 73–76.

8. Augustus W. Greely, *Handbook of Polar Discoveries* (Boston, MA: Little, Brown, 1909), pp. 189–90.

9. Quoted in John Edward Weems, *Race for the Pole* (New York: Henry Holt, 1960), p. 30.

10. Coverage of Peary's expeditions is based on ibid.; Caswell, *Arctic Frontiers*; Jeanette Mirsky, *To the Arctic* (Chicago, IL: University of Chicago Press, 1970); Markham, *The Lands of Silence*; J. Gordon Hayes, *The Conquest of the North Pole* (New York: The Macmillan Company, 1934); and Mowat, *Polar Passion*.

11. Donald B. MacMillan, *How Peary Reached the Pole* (Boston, MA: Houghton, Mifflin, 1934), pp. 128–30, 157–58, 172–73. Stefansson's comments on Peary's use of Eskimo techniques also are of interest. They appear in *The Friendly Arctic: The Story of Five Years in Polar Regions* (New York: The Macmillan Company, 1943), pp. 128–29, 135. See also Robert Edwin Peary, *Secrets of Polar Travel* (New York: Century, 1917).

12. Discussions of Peary's navigation appear in MacMillan, *How Peary Reached the North Pole*, pp. 283, 291; and Markham, *Lands of Silence,* p. 357.

13. See, for example, MacMillan, *How Peary Reached the North Pole*, p. 291.

14. Accounts of the Cook expedition appear in Mowat, *The Polar Passion,* pp. 237–84; Theon Wright, *The Big Nail: The Story of the Cook-Peary Feud* (New York: John Day, 1970); and Hugh Eames, *Winner Lose All: Dr. Cook and the Theft of the North Pole* (Boston, MA: Little, Brown, 1973). Markham, *Lands of Silence*, pp. 354–55 offers critical comments on Cook's navigation.

15. Quoted in MacMillan, *How Peary Reached the North Pole,* p. 283.

16. Enumerable volumes discuss this controversy. Among those favoring Cook or opposing Peary are Hayes, *The Conquest of the North Pole*; Mowat, *Polar Passion*; Wright, *The Big Nail*; and Eames, *Winner Lose All*. Examples of pro-Peary books are L. P. Kirwan, *A History of Polar Exploration* (New York: W. W. Norton, 1959); Mirsky, *To the Arctic*; and Weems, *Race for the Pole*.

17. Mowat, *Polar Passion,* pp. 287–88.

18. Paul-Emile Victor, *Man and the Conquest of the Poles* (New York: Simon and Schuster, 1963), pp. 278–82; John C. Reed, "United States Arctic Exploration Since 1939," in Herman R. Friis, ed., *United States Polar Exploration* (Athens, OH: Ohio University Press, 1970), pp. 26–28; John C. Reed and Andreas G. Ronhovde, *Arctic Laboratory* (Washington, DC: The Arctic Institute of North America, 1971), passim.

19. See account in *The New York Times*, 26 January 1969, p. 26.

20. *The New York Times*, 2 May 1986, pp. 1, 36.

21. Virgil F. Keith, "Across the Top," U.S. Naval Institute *Proceedings*, August 1970, pp. 60–69.

22. William R. Anderson with Clay Blair Jr., *Nautilus 90 North* (Cleveland: World Publishing Company, 1959) is the basic account.

23. Excellent coverage of this operation appears in James Calvert, *Surface at the Pole* (New York: Simon and Schuster, 1963).

24. George P. Steele, *Seadragon: Northwest Under the Ice* (New York: G. P. Dutton, 1962).

15 "FLYING OVER THE POLAR SEA"

LCDR Richard E. Byrd, USN

In May 1926 Lieutenant Commander Richard E. Byrd made a polar flight that he claimed reached the North Pole, and it made him a national hero. That flight would become the subject of controversy, and Byrd would continue polar exploration for the U.S. Navy with two major Antarctic expeditions. However, before all this, from June until October 1925, Byrd commanded the aviation unit of the Arctic explorer Donald B. MacMillan's North Greenland expedition, which was backed by the National Geographic Society. Byrd's article was originally published in August 1925, during that expedition. It is a vivid reminder that U.S. Navy interest and operations in the region are not new and are, in many ways, just as challenging.

"FLYING OVER THE POLAR SEA"

By LCDR Richard E. Byrd, USN, U.S. Naval Institute *Proceedings* (August 1925): 1319–39.

Editor's Note

The following article was written by Lieutenant Commander Byrd on board ship while making passage from Wiscasset, Maine, to Sydney, Nova Scotia, and mailed to the U.S. Naval Institute just prior to the departure of the expedition from Sydney for Etah. It is, therefore, the last written account of the proposed flying operations over the Polar Sea by a member of the expedition. This article, although written exclusively for the *Proceedings*, was made possible by the courtesy of the National Geographic Society.

When Admiral Robert E. Peary, in 1906, made his farthest westward trip to Axel Heiberg Land and the western part of Grant Land, he twice thought he saw from the shores of the Polar Sea the summits of a high land. The first time he saw it was from the top of Cape Colgate, 2,000 feet above the sea level. Of this experience Peary wrote: "North stretched the well-known ragged surface of the polar pack, and northwest, it was with a thrill that my glasses revealed the faint white summits of a distant land which my eskimos claimed to have seen as we came along from the last camp." Again, he saw this land from Cape Thomas Hubbard. "I could make out," he wrote, "apparently a little more distantly the snow-clad summits of the distant land in the northwest, above the ice horizon." Peary called this land Crocker Land. Eight years later Dr. Donald MacMillan went out 150 miles over the Polar Sea toward that land but he never reached it. In his book he refers as follows to his experience one hundred miles out from the shore: "There could be no doubt about it. Great heavens! What a land! Hills, valleys, snowcapped peaks extending through at least 120 degrees of the horizon." He went on fifty miles further and the land got no closer, so he decided that it was a mirage and mirage it seems to be.

But, between Point Barrow, Cape Columbia and the North Pole there lie one million square miles of unexplored region and there are indications that within that area there is land. For example, at Point Barrow, Alaska, the flood stream of the tide should come from the north if this million square miles were but a deep polar basin. The times and ranges of the semi-daily tide from the Greenland coast to the Alaskan coast also indicate that there is land in the Polar Sea. Another indication is the great age of the ice found in Beaufort Sea.

At any rate, the mission of the Naval Arctic Aviation Unit is to co-operate with the MacMillan expedition in the exploration of some of this vast region. The expedition is going under the auspices of the greatest geographic society in the world, the National Geographic Society which publishes the *National*

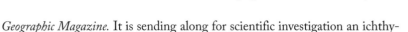

Geographic Magazine. It is sending along for scientific investigation an ichthyologist and a geologist and the naval unit includes in its personnel an aerologist.

Types of Planes Used

The Navy has assigned three Loening amphibian airplanes to this expedition. They have been designated the *NA-1, NA-2,* and *NA-3.* It is a new type of plane designed and built by Mr. Grover Loening. Only four of this particular type have been turned out. The Army got the first one and would have received the next three had not General Patrick generously given the Navy priority of delivery on them for this Arctic expedition. In order to get the propeller forward on this amphibian it became necessary, due to a protruding pontoon, to keep it high by inverting the Liberty motor. It is interesting to note that on account of the cooling system the inverted Liberty develops a little more power than the upright Liberty.

The wheels are raised and lowered by an electric motor and they can also be moved by a hand crank, in case the battery gives out. The pontoon extends several feet beyond the propeller and makes a satisfactory seaplane when the wheels are up. The plane is practically all metal with the exception of the fuselage bracing and the wing covering. The whole machine is compact and strongly built. Its wing spread is forty-five feet and length thirty-three feet. It is a two-place machine with controls both in the forward and after cockpits.

With an extra gasoline tank, two pilots, navigation equipment, rifle and ammunition, and a month and a half supply of food aboard, she is capable of making about one thousand miles if the take-off is from the land. That is a splendid performance for an amphibian. With the same equipment aboard, she could take off from the water with sufficient fuel for about seven hundred miles flight.

Bases

The S.S. *Peary* and S.S. *Bowdoin,* which are conveying the amphibians and the pilots and mechanics into the Arctic, will base somewhere near Etah, 713 miles from the North Pole. There is no land landing field at Etah. It will, therefore, be necessary to look up and down the Greenland coast until a suitable field is found. It would be dangerous to base the planes in the water, due to the drifting ice in Smith's Sound. There are periods when the water may be clear of ice, but no one can predict when the ice will begin to descend. The planes would be crushed in between the floes.

The *Peary* is only 134 feet long and there was some difficulty in getting the planes and equipment aboard. The planes are stowed astern, close together, with the wings removed. They cannot be erected on the ship, due to lack of space, and it would be very difficult to put the wings on in the water, not to speak of the danger of drifting ice. An endeavor, therefore, will be made to find a beach so that two wingless planes can be put overboard, lashed together and towed ashore. Without the wings or other support, a plane in the water would, of course, upset. When the beach is reached the wheels could be lowered and the planes pulled onto the land by their propellers. There is such a beach twenty miles south of Etah. With the two planes off the ship the wings could be placed on the third plane on the ship and she could take off from the water and land at the land base. The advantage of the amphibious quality of these planes is clear.

After the first base is formed on land, practice flights will be made for the following purposes:

1. To give each motor ten or fifteen hours flying (including what they have already had), so as to get as many as possible of the kinks out of it before the actual expedition is started. There is no reason why the inverted Liberty should not be as reliable as the upright Liberty, but, like all new things, it has had many kinks to overcome. The Naval Aircraft Factory took out most of these troubles, which, in the short time allotted, was very remarkable. The factory worked day and night, and had the Navy had no such institution available, this polar project would have been impossible this year.

2. To test out radio communication and radio compass bearings.

3. To investigate landing conditions toward Axel Heiberg Land and Cape Columbia. Cape Columbia is to the northward of Etah and is 413 miles from the pole. To go there the flying would be over water and ice and there will probably be landing places along the route in

the water in the inlets along the irregular coast lines. To get to Axel Heiberg Land there will be rugged land and some fjords to fly over. It is thought that the base on the shores of the Polar Sea probably will be formed at one of the above-mentioned places—the one that affords the best landing facilities en route. We know that there are landing places at Cape Columbia and Cape Thomas Hubbard, Axel Heiberg Land.

It would not seem wise to load the planes to the limit for the character of work that lies ahead of the naval unit because to make the proper speed the motors would have to be opened up to such an extent as would lessen their reliability. Therefore, the most practical way to accomplish the mission of the unit is to form an *intermediate* base *half way* between the *surface ships* and *the base* on the Polar Sea: in other words, to work gradually toward the objective rather than make spectacular dashes with overloads.

The distances from Etah (the ship base) to the following points are approximately as follows: to the pole, seven hundred miles; to Cape Columbia or Cape Hubbard, three hundred miles; therefore it is seen that if the naval plane unit succeeds in establishing a base on the Polar Sea, and an intermediate base between it and the ship base (Etah), the planes will at all times be not more than seventy-five miles from some one of the three bases while the preparatory work is being carried out before flights are made over the Polar Sea. At each of these bases will be food, engine fuel, supplies and fire arms. The base on the Polar Sea should have, in addition to the above: a radio operator, radio outfit, one eskimo, sleeping bags and one small tent. The success of the establishment of such an intermediate base should have military significance, in that it will have been demonstrated that small plane bases can be located by amphibian planes.

Radio

Every effort has been made by the Navy to insure getting two-way communication from plane to plane and between the planes and the two bases—the ships and the base on the Polar Sea. The planes now have installed on them both high and low frequency, the former supplied by the Zenith Radio Corporation of Chicago and the latter, standard navy sets. For the flights out over the Polar Sea one plane probably will be equipped with a high frequency set and the other with the low frequency. The third plane will be held in reserve at the Polar Sea base. This plane, of course, will also be equipped with radio. The radio set at the Polar Sea base will not be powerful enough to transmit radio bearings direct to the planes, so they will be sent to the base ships (Etah) which will relay to the planes their bearings from the Polar Sea base. It is doubtful that entire dependence for navigating can be placed upon these bearings. The high frequency set has the advantage that if down and not smashed the plane can still send messages back to the bases. Though the low frequency sets cannot send from the ground they have the advantage of being standard equipment and, hence, thoroughly known.

It is common knowledge that it is very difficult to get radio messages through during daylight. Of course, the naval unit will be operating in the Arctic during that time of year when the sun does not set. As high frequency sets are the only ones that can get through during daylight the expedition has been supplied with them and it is hoped that the *Peary* and the *Bowdoin* can at all times keep in touch with the Navy Department. That, however, cannot be guaranteed.

Navigation

Navigating a plane over the North Polar Sea presents some interesting problems. The Magnetic Pole lies on Boothia Peninsula about twelve hundred miles south of the geographic pole and the part of the unexplored region over which the naval unit expects to fly lies between the two poles. Consequently, variations as high as 180 degrees will be encountered. The exact amount of this variation, however, cannot be known except by actual observations. The Navy's Hydrographic Office has constructed a "Line of Variation" chart of the polar regions for the naval unit, but these theoretical values may be in error as much as ten or fifteen degrees.

To add to the difficulties of the dead-reckoning navigation the ordinary magnetic compass will probably not be dependable in a plane, due to the weakness of the horizontal component of the earth's magnetism in that region. For example, Washington, D.C., has a seven times greater component of horizontal force than there is in the region over which the naval unit is to fly. There is more of a tendency, of course, for the compass needle to swing in an airplane than there is in a surface ship. One of Amundsen's aides reported that the ordinary magnetic compass was of little use even as far from the Magnetic Pole as Spitzbergen. However, this question can be settled only by actual trial.

It would have been unwise, therefore, to take airplanes into the Arctic putting entire dependence upon the magnetic compass. A sun pelorus made by Kelvin and Wilfrid O. White Company has been procured. This instrument is set for the local apparent time, declination, and latitude of a place, and the true course is shown on a dial by a shadow thrown by a shadow pin. To steer a course with this instrument, however, would require calculations and a resetting of it at intervals, due to the changing azimuth of the sun. It became advisable, then, to construct a similar instrument with a twenty-four-hour clock so that the shadow could be made to follow the hands of the clock and do away with the necessity of repeatedly resetting the dials. Mr. Bumstead, of the National Geographic Society, was requested to construct this compass, which he did in a very short time, working day and night. The first instrument was successful and Dr. Grosvenor then took him off from all work for the society so that he could give his exclusive time to constructing two more sun compasses. These compasses were delivered the day the expedition reached Wiscasset en route to Etah. One of them was tested by taking it with a navy standard compass to a locality free from local metals, setting it to north, and moving the face of the watch parallel to the equator, setting the twenty-four hour watch to local apparent time, and then comparing the compass with the true north thus indicated. (The sun compass shows the true course it is set to when the shadow thrown by the shadow pin falls on the hand of the watch.) The difference between the north of the compass and the north of the sun compass was exactly the amount of the variation in that locality. There will be errors when the plane is not flying level but the general direction can be followed.

Of course, the Mercator projection is not practicable in the polar regions. The most suitable chart available is a conventional projection, the polar chart,

No. 2560, published by the Hydrographic Office of the Navy Department. On the polar chart the rhumb line would be spiral but the spiral would not become very ominous until very near the geographical pole. When steering by the magnetic compass alone (when between the magnetic and geographical poles) it would seem advisable to use the rhumb line, the curved course on the chart, because to steer a straight line would necessitate changing the true course and the variation every few minutes, whereas the rhumb line would necessitate only changes of five degrees in variation every fifteen minutes. If the navigator is attempting to steer a straight line and gets off his course because of fog or unknown variations or bad compasses or other causes, he will then be in a bad shape for he will be applying the wrong corrections when he changes his variation and his true course; his errors will be cumulative and he may readily get lost.

If Amundsen's reports so far published are correct, they indicate very clearly that he was considerably off his course and wasted much gasoline because of it. He was steering due north, so his true course should have remained constant and his variation between Spitzbergen and the pole should have changed only ten degrees. But, when the fog (and certainly it was no fault of his), got him off his course he began cutting lines of variation and his troubles began. On his return trip he landed one hundred miles to the eastward of his course.

It seems that the best method of using the sun compass is to set it for the local apparent time and latitude of the middle meridian that is to be passed. This gives for all practical purposes a great circle which can be drawn as a straight line and it will not be necessary to apply so many changes of uncertain accuracy. An approximate rhumb line, however, can be steered with the sun compass by resetting the course dial every twenty miles or so. Science has no information on variation in the regions to be flown over; it is hoped that the use of the sun compass will facilitate the obtaining of information concerning this uncertain quantity. Of course, when the sun is behind the clouds the sun compass is useless and other means had to be found to supplant the sun compass and the magnetic compass in case the occasion should arise.

The Pioneer Instrument Company has developed an earth induction compass which is much more sensitive than the age-old magnetic compass and will probably react to the small horizontal component of the earth's magnetism that prevails in the Arctic regions. It is subject, however, to the same errors of large and unknown variation as is the magnetic compass. Its principle is simple. There is a coil which is revolved by a small electric motor. When this coil revolves in the field of the earth's magnetism a current is generated which actuates a pointer on the pilot's instrument board. When the instrument is set to the desired course (after applying the variation and the deviation), the pointer is on zero and the coil is parallel to the lines of force of the earth's magnetism. Any movement of the plane off this course turns the revolving coil at an angle to these lines of force and so induces a current of electricity which increases with the angle the plane is off the course. This throws the steering pointer off zero and it becomes the pilot's job to bring it back to zero. This compass is subject to a deviation which is especially large when the plane is not flying on an even keel. Other deviations can be eliminated by placing the coil far out on the wing where it will not be affected by the metal, switches, radio, etc., of the plane. The standard navy steering compass in the forward cockpit of these amphibians showed some deviations as high as eighty degrees. It became advisable, then, to put another compass in the after cockpit away from the starting switches.

Astronomical Observations

It is evident that quick and reliable methods of locating the Sumner line must be available to the naval unit. Mr. G. W. Littlehales, the Navy's hydrographic engineer, with his usual genius, filled in the breach by calling attention to the very interesting situation peculiar to the spot that we call the North Pole. The north pole of the heavens and the zenith coincide. Hour circles and meridians of longitude coincide with azimuth circles or vertical circles. Parallels of declination coincide with parallels of altitude. That means that the values of declination of the sun and moon as given in the *Nautical Almanac* for any given time equal the altitudes of those bodies at the North Pole. The moon will be visible in the Arctic regions during August. The observer taking sights in the Arctic regions can take the North Pole as his computed point and he has his altitude for that point all worked out for him in the *Nautical Almanac*. But that may better be left to be told by Mr. Littlehales himself.

The naval unit is provided with sidereal watches as well as G.M.T. clocks in order to facilitate working out moon sights. In the location where the flying is to be done, using the above-mentioned hydrographic chart, it will be sufficiently accurate to lay down the Sumner line as a straight line on the chart so that the lines of position can be plotted graphically with almost no calculation.

The following instruments developed for the Navy's transatlantic flight (1919), and which are now standard equipment, will be used: bubble sextant, speed and drift indicator, course and distance indicator, and smoke bombs, the latter being an improved type. The drift indicator should show (when not in fog) the drift within two degrees. This is a big factor in air dead-reckoning for it sometimes greatly changes the course and speed. The course and distance indicator is an instrument which solves the triangle of forces without resort to mathematics. Knowing the speed through the air and the drift angle, it is possible to find the speed over the ground by use of this instrument.

Another method that will be utilized in finding the drift is effected by having radial lines drawn on the horizontal stabilizer, five degrees apart, with the position of the observer's eye as a center. By dropping a smoke bomb and noting on the graduated radial lines the angle at which the smoke shows up aft the angle of drift is ascertained.

Food

This article would not be complete without mention of the food to be carried to the base on the Polar Sea. Lieutenant Commander J. R. Phelps, Medical Corps, U.S. Navy, has made an exhaustive study of food values with special consideration given to the great limitation of weight-carrying in aircraft. The object was to supply the proper proportions of proteins, fat, carbohydrates, and vitamins needed by the human body to make possible a carefully balanced diet with regard to actual experience over a period of many years in the Navy. The great Arctic food is pemmican, which consists of 53 per cent dried beef, 31.5 per cent oleo oil, 9.5 per cent raisins, and 6 per cent sugar. Lieutenant Commander Phelps has worked out a diet that is undoubtedly much superior to the pemmican diet, though pemmican is included. He concluded that five thousand calories per man per day would be needed for the work in hand.

Two months' food will be taken to the base on the Polar Sea for the seven or eight men who will be there. Someone else might have approached this problem in a somewhat different manner. Different articles of food might have been recommended. But, the assumed variables with which we must deal necessarily limit selection to a few outstanding foods. The caloric value per pound must be high and the foods chosen must be acceptable to those who will eat them. It is necessary to assume that little or no cooking may be possible and that you may have to fall back on high protein and high fat concentration in case of accident involving a stay through one winter. At the same time, provision should be made to furnish a list of foods that will permit a reasonably varied diet consisting of cooked food served hot, in case everything goes well.

16 "REAR ADMIRAL PEARY, U.S.N., SCIENTIST AND EXPLORER"

LCDR Fitzhugh Green, USN

The polar explorer Robert Peary was sixty-three when he died in 1920. His claim of reaching the North Pole on April 6, 1909, during his final 1908–9 expedition, has long been the subject of controversy. This glowing reminiscence was written two years after Peary's death by Lieutenant Commander Fitzhugh Green, a friend of Peary and a fellow naval officer and explorer who knew many of the polar explorers of the era. It does not address the controversies that surrounded Peary. However, it does discuss the challenges faced by any naval leader operating in the Arctic. Writing of the challenges of leadership and operations in the Arctic, Green, a colorful personality himself, notes: "Arctic work magnifies the leader's difficulties a thousand times."

"REAR ADMIRAL PEARY, U.S.N., SCIENTIST AND EXPLORER"

By LCDR Fitzhugh Green, USN, U.S. Naval Institute *Proceedings* (August 1922): 1315–24.

Editor's Note

Lieutenant Commander Fitzhugh Green is well qualified to write a sketch of Admiral Peary's life. Lieutenant Commander Green has always been a hunting and camping enthusiast. Study of arctic literature and a close friendship with Admiral Peary eventually led him [Green] to join the Crocker Land Expedition. He spent over three years with the Smith Sound Eskimos in North Greenland and Ellsemere Land. In 1914, Green and Macmillan established a new record to the Northwest on the Polar Sea. The author of this article has enjoyed the friendship of Amundsen, Shackleton, Stefansson, Evans, Rasmussen, and others in our generation who have with untiring zeal helped blaze the long white trail to the Ends of the Earth.

Robert E. Peary was born in Cresson, Pa., May 6, 1856. He passed his boyhood in Maine and graduated from Bowdoin College in 1877. He entered the Coast and Geodetic Survey from which he joined the navy as a civil engineer in 1881. By 1885, though only a lieutenant, he was engineer-in-chief of the Nicaraguan survey. He invented a new type of lock-gates for the proposed canal and otherwise distinguished himself from the common run of youthful pioneers.

His Danish friend Maigaard literally broke the ice for the second period of Peary's life. He persuaded the young officer to accompany him to Disco Bay, Greenland, for an ice-cap reconnaissance in 1886. Instantly and deeply was Peary bitten by the exploring bug. And nothing could have demonstrated his personality and temperament better than that he threw all precedent and centuries of experience to the winds: he determined to strike out on a new unbeaten path to the Pole.

Since Hakluyt's day men had tried to penetrate the Polar regions by water routes. There were three main gateways:

1. North around Scandinavia past Spitzbergen.
2. Through Baffin Bay up the Greenland coast.
3. By Alaska into Behring [Bering] Sea.

Early Europeans used the first route because it enabled them to cling to their continent. As the second, by Greenland, meant crossing the North Atlantic it did not become popular until in the last generation when both British and American expeditions raced it neck and neck. The third one, of which the take-off is Point Barrow, Alaska, never has been conquered even to a mild degree.

Such tragedies as De Long's voyage in the *Jeannette*, our own naval enterprise, serve only to emphasize the desperate nature of tackling an untried arctic route.

But this is what Peary did. Nor was it simply a case of laying his ship on a different course, or threading his way through another sound or channel as other courageous explorers have. His plan was infinitely more daring. He chose the Greenland ice-cap.

Greenland is a great pear-shaped continent nearly 1,500 miles long, and something like 900 miles broad across its upper bulge. It is the classic example of the glacial age. Except for its rocky fringes and the southern tip, it is buried in ice. Depth of this frosting is conjectural. Along the coast 2,000-foot cliffs are literally dwarfed by the ice dome rearing back of them more than 9,000 feet into the sky.

To seek the Pole by the Greenland ice-cap could be compared to hunting a wild elephant by crawling up its back until close enough to shoot it behind the ear. A wild elephant's back is a tame comparison to that blizzard-tortured desolation Peary chose as his way north.

For practically ten years he fought a losing battle. True, he learned the technique of arctic travel from the natives; how to drive dogs; how to build snow igloos; how to exist on blubber and raw meat; how to wear skin clothes; how to hunt without firearms; how to sleep sitting up without a sleeping bag; how to burrow into a drift when taken unaware by a howling blizzard. He brought back three ponderous meteorites, one of them weighing ninety tons, the largest known to man. He collected a large quantity of information of scientific value as well as of popular interest. But so far as his ambition to reach the Pole was concerned he had failed.

Not until 1898 did he abandon the ice-cap. Old-timers said, "I told you so," when he admitted the Pole could not be reached that way. Also, they pointed out the foolish enthusiasm of youth and inexperience. Indeed, the injustice of the criticism Peary suffered at this critical stage of his life would have broken a weaker character.

He was forty now. He had battered away at a problem the geographical world declared could not be solved by the means he chose. That he had determined the insularity of Greenland was largely ignored. That he had made several ice-cap trips of over 1,300 miles, driving his own sledges and with no help save that of his tenderfoot companions, was not advertised. Yet Nansen, his contemporary, had secured a place of exalted fame from one summer's crossing of the narrow southern tip of Greenland. Nansen had food and Lapps and no back trail to figure on. Peary did five times the distance on little more than raw walrus meat and blubber. Eskimos helped him up the first glacier, then deserted him in terror at the prospect of travelling the great unknown interior.

The second period of Peary's northern work began in 1898 when he based on the opposite side of Smith Sound. By this time he had won the confidence of the natives. He realized the value of their dogs and meat. Greely, Kane, Hayes, and all the British expeditions up this coast had deprecated the Eskimo's value as an adjunct to arctic work. Peary cleverly recognized that no white man can pretend to lead his natural life in high latitudes. He trained himself and his men to be Eskimos. He went into their igloos and became one of them. He learned to eat their food, to wear their skin clothes, to perform the innumerable little tricks of keeping warm and dry that may mean the difference between life and death when caught in a sudden crushing snow-filled gale. When he crossed the Ellsemere Land he took enough families with him to provide hunters, seamstresses, and dog drivers.

It must be recalled that the upper left-hand (western) corner of Greenland converges with the land opposite. Arctic currents drag a jam of ice through this narrow passage. Open water is almost unknown. For this reason expeditions have usually disembarked in the Smith Sound region about 78° north latitude, and made the rest of their way by sledge. Greely, Kane, Nares, and one or two others have happened to find open years which permitted them to base further north than Peary's earlier attempts. But their terrible loss of life and suffering offset any advantage they gained in the beginning.

Now began a new kind of work for Peary. On the ice-cap, despite its frightful wind and cold, smooth going was the rule. Coastal sledging means one unending struggle over the mountainous ice-jams that are formed by tide and current. On the ice-cap Peary had simply to grit his teeth and plug away. So long as he didn't starve or freeze, and held his course, all he had to do was to keep on lifting one foot after the other in order to reach his destination. Along the Ellsemere coast ten minutes of travelling on the level was a luxury that was stimulating. The average day consisted of a kind of exaggerated football game. Five hundred pound sledge load was the ball. Each driver carried, drove, pushed, and pulled his own. The opposing line was a serrated ridge of huge sharp ice fragments from ten to fifty feet high. And when by sheer strength and ice-axe this line was passed, there was always another just beyond. The enemy's tactics varied by cunningly placed lanes of open water, the most dangerous trap in the north.

It was a hard school, and Peary was a pupil for another ten years. It is interesting that this corresponds to the time he spent settling ice-cap delusion. But he was learning every day of it. His troubles came so fast that his publisher never gave him room to list them. On the Greenland coast he had only crushed his ankle on shipboard, lost a year's cache of stores on the ice-cap, and the like. On the Ellsemere coast he froze off every toe except a fraction of each large one. At Cape Sabine his Eskimos died like flies from a kind of dysentery. Only a few years ago the writer found the bodies of two of them wedged down between the boulders. Clothing had not even been removed from the mummified remains. Peary's white companions even pronounced him insane for the seemingly super-human campaigns he planned.

One misfortune after another assailed him at home. He was told that he would receive no more leave from the Navy Department. What this meant can be appreciated only in the light of the ten years' struggle he had already endured for the sake of his ambition. His finances, never plentiful, ceased to exist. Only by dint of lecturing and writing to a degree nearly as exhausting as his travels over northern trails was he able to accumulate the pitifully scanty store of supplies and equipment he carried on these earlier expeditions.

These were the blackest years of his life. Earlier he had been able to exploit the novelty of his schemes as a means to further successive expeditions. The meteorites, the crossings of Greenland, and the like were stepping-stones to that notoriety so profitable to the lecturer and writer. Now the novelty had worn off. In despair he wrote to a friend:

"I seem to have reached the end of the lane. What I gained in Greenland I am losing in Ellsemere Land. Soon they will call me a fake, say I am just hoarding what I can mint from the public's credulity. Yet, would it profit me to confess that all my savings, all of my wife's savings, all I can beg, borrow, or steal, have gone into the quite inadequate equipment I have gathered for the next attempt?"

Such indomitableness is worthy of the best traditions of the navy.

It is difficult to realize the abysmal discouragements that sprang up on every hand. Members of earlier expeditions, broken by the inhuman toil of arctic sledging and embittered by the petty quarrels isolation breeds, lost no opportunity to backbite and discredit the man who they jealously realized was a better man than they. He dared to go again. Eskimos behaved with childish petulance when he returned without the promised gifts his anemic capital sometimes couldn't buy. And constant worry prevented his training like an athlete for the physical strain of his long exhausting trips. Instead, his naturally rugged constitution was sapped of its reserve endurance. Inanition was the logical result; and on practically all his later sledging Peary was a victim of what northern travelers call "starvation bloat." Formerly this distemper was erroneously thought to be scurvy. It is painful swelling of limbs consequent upon overwork and under feeding. Peary himself has described his sensations when suffering from it as the combination of "elephantiasis and toothache"! No more agonizing climax to his troubles could be conceived.

A flood of friendship turned the tide. Peary's magnificent tenacity had finally won him the attention and interest of a group of influential New Yorkers. When the Navy Department flatly refused to even consider further leave to prosecute more northern work, Mr. Charles Moore personally convinced President McKinley that the public would hold our government gravely culpable for cramping the energies of the one man who might redeem American geographical prestige so marred by recent arctic fiascoes. When burden of debt was about to place both Peary and his wife in legal toils Mr. Morris K. Jesup and a score of others reached down into their own pockets and produced. When the ordinary run of vessels fit for ice work still fell short of the strength required to face the polar pack, the *Roosevelt* was built. Peary's own designs were used.

The third period of Peary's life was now on. In the first he had made his ice-cap struggle. In the second he had crawled up and down the Ellsemere coast. He now launched his startling scheme to force a ship to the very shores of the Polar Sea. By basing there and with the help of dogs and Eskimos he might well hope to make the 500-mile dash to the pole and get back before spring leads cut him off. Only his intimate knowledge of ice and current conditions north of Cape Sabine made this plan in the least way possible. It was a long known fact that the very passage Peary selected was the most ice-choked hole north of the arctic circle. He had observed year after year that under certain combinations of wind and tide acting over limited combinations of shoal and deep, the ice jam suffered itself to be ripped and split until slender black leads of open water strung here and there between the Polar Sea and Smith Sound. Peary determined to gamble on the position of these leads. If the ship were caught she might suffer the fate of the *Proteus* which sank in eighteen minutes after being nipped. If he succeeded in getting through, safe return was even less likely.

He got through and he got back. He began to break records. In 1902 he made a new American record with 84° 17′ north. In 1906 the world's record was smashed when he reached 87° 06′ north. On April 6, 1909, he reached the Pole.

This was a many-sided triumph. Obstacles at home had been overcome. Ice navigation had been set a new standard by the *Roosevelt*'s performance. But beyond a shadow of a doubt the greatest factor in Peary's success was the remarkable sledging technique he had developed in twenty-five years of practice.

He worked on the unit basis. Logistics of his units were as perfect as mathematical reason could make them. A unit comprised one man, one sledge, and one team of dogs. This unit was independent and self-supporting for fifty days. The sledge carried about 650 pounds to start with. Of this 500 pounds of dog pemmican gave each animal one pound per day. Fifty pounds of hard biscuit and fifty pounds of man pemmican gave the driver a pound of each a day. A few gallons of kerosene, a tiny stove, tinned milk, tea, and some extra footgear and mittens completed the outfit.

A snow igloo will hold four men comfortably. A gallon of tea at night and another in the morning will keep four men going. Four men can just about handle one 600-pound mass over a seventy-foot ice pressure-ridge. One white man can keep three Eskimos busy, amused, fed, cheered, and loyal. Thus four men, one white and three brown, was the size of party Peary chose to group his helpers in.

Dogs and men have one characteristic in common, both easily and gladly follow where another has gone. This quality is an instinctive physical and mental laziness. For men or dogs to break their own trail is toil. It means concentration every moment to avoid pitfalls and rough spots. It means picking footholds. It means a kind of subconscious grief and jealousy over the ease with which those behind must come.

So members of each party of four took their turns at trail breaking. And when the whole party had been exhausted by a week or so of battering down ice walls, cutting ice roads, and plowing heavy snow, it stood aside and let the next party, fresh and eager, take the van.

This plan made it possible to line the trail with food. The advance patrol wore out physically before it had used all its supplies. It cached what it had, turned, and sped back light, leaving its extras for those to come. In this way physical energy of men and dogs was conserved. Land advance in the recent war was made by the same system of trail breaking and replacements that was used on the Polar Sea.

It is not difficult then to understand how Peary placed himself with fresh dogs, picked men, and adequate supplies within easy striking distance of the Pole. Conversely, it is not conceivable that a single unsupported sledge party could make the entire trip out and back as Cook claimed to have done. Peary's strongest trail breaking party collapsed at one quarter of the distance to be covered. Six groups were thus expended in blazing the way.

Peary has been scored for the un-seagoing nature of his work. This and lack of affiliation with naval men necessitated by his long absences no doubt account for present professional disregard of his achievements. Even the Navy Department indicated grave doubt as to the ultimate value of his oceanic and other scientific investigations. And there existed an undeniable feeling that every year he spent north took him that much further away from nautical skill of any sort or degree.

No view could be more unjust. Few naval officers have had to face afloat the gauntlet of perils and difficulties Peary ran on every voyage north. His personnel consisted of scientific assistants wholly unversed in sea matters, and a motley crew of "down-east" fishermen. Better classes of deck-hands and engine-room force would not sail away for an indefinite period on the meager wages Peary was forced to pay.

He had but one real aid, Bob Bartlett, who acted as a kind of executive officer so far as the ship was concerned. The two of them took the *Roosevelt* through thousands of miles of uncharted, unlighted, ice-filled seas. When the rudder was bitten off in an ice-jam they beached her and rigged a new one. When fuel ran out they dragged into a Labrador fiord and cut driftwood.

To illustrate his seamanship and his engineering ability, take one meteorite he brought back which weighed over ninety tons. The mass was well back in the hills. There was no dock to handle it from shore to ship. Time was painfully limited by constant snow storms, gales, and ice. Yet Peary wrestled the huge lump down, got it aboard his little vessel, and turned it over to the 100-ton crane in the New York navy yard without a hint that he was doing more than his normal duty. Compare this absurd equipment with that used at the gun factory in handling a sixteen-inch gun or a weight similar to that of the meteorite, and some idea of the feat may be formed.

One remarkable aspect of Peary was his extraordinary veracity. A queer implication, no doubt. But in going back over records of other explorers it is almost impossible to find one that did not consciously or unconsciously exaggerate. It is so easy to create a false impression among those at home who are not familiar with arctic work. It is so safe and simple to fill in one's tale with details that make it vivid and alluring, yet which are no more true than the narrative of a novelist who knows his atmosphere and manufactures his episodes to fit. Peary never succumbed to this temptation. When he had gone his sledging limit he took the best sight he could, made the best notes his weary brain would permit, built the highest cairn his exhausted muscles could manage, and came home. This statement has been proved literally thousands of times. To the popular mind it means little; the popular mind loves to be duped. By the naval officer's standard there is no middle ground between truth and untruth. But to the world of science there is a middle ground, and it is by far the widest area of all research and information. Peary was the single explorer in the history of the north who entirely avoided it.

This same sense of truth and justice may have accounted for his ability to handle the varied elements of his subordinates. Eskimos are childlike. They are brave, devoted, trustworthy, and honest one moment; the next they may be just the opposite. Superstition can turn their courage to terror in an instant. Pique alone is sufficient to bring about desertion. Honor among most primitive races is a matter of convenience. To keep the services of such people through all the years he did, was a monument to Peary's tact and diplomacy.

Then there were his scientific assistants. The north can poison a white man's mind. Arctic neurasthenia is a common malady among Danish in South Greenland. Few expeditions have escaped insanity in some form or other. When the leader portions out skins and food and fuel he may be measuring the life span of his men. Fuel and food can be allotted with geometrical nicety. Skins and Eskimo assistants, routes, weather, and the like cannot. The white driver fights a blizzard, freezes, starves, thirsts, and sometimes dies, all the while cursing his leader for the fractional partiality he seemed to have shown in assigning garments or the course of this particular trip. Peary's command used to return hating him in a way that murder couldn't gratify. Every arctic leader has had the same experience. When the antidote of normal life neutralizes the poison, this bitterness is gradually dissolved. But by that time all are separated and again engaged at home in their individual pursuits. While north, when friendly spirit is the one human bond, only the strongest character can rise above the terrible mental depression and irritableness which isolation and physical suffering bring. Peary could; he had the incentive. But those with him year after year could not. Only by magnificent patience and forbearance was he able to hold his authority over them with sufficient strength to utilize their aid to the end.

As navy men we cannot but admire such supreme qualities of command. To be sure aboard ship we have our uniforms and constant touch with Washington, routine, drill, and discipline to remind us of the authority vested in our superior officer. But even with all this there must be something more. There must originate in our leaders a spirit of professional cohesion which will permeate the entire ship's company and imbue them with a unity of purpose and ideal. This spirit must be strong enough to rise above discomfort, monotony of task, and every form of temperamental incompatibility. It is the priceless virtue which every officer spends his life in learning to master and to use. Yet, alas, how few are truly successful.

Arctic work magnifies the leader's difficulties a thousand times. Peary, for instance, had two masters and three groups of subordinates. He was answerable to the Navy Department and to his scientific backers; he had to direct and use successfully, first, a tribe of Eskimos; second, an always unruly crew; third, his special aids who were almost 100 per cent tenderfeet on every trip he took. His methods under the circumstances warrant the study of every one of us. His success deserves our most genuine pride. And the heroic tenacity of purpose that finally won him the goal of his life's endeavor, that took his life in the end, must be justly added where it belongs—to our capital of traditional service valor.

17 "THE ARCTIC OCEAN—HOT TIMES IN A COLD PLACE"

Don Walsh

In this essay Don Walsh provides an overview of the properties of the Arctic Ocean and gives a succinct description of potential oceanographic conditions for future maritime and naval operations above the Arctic Circle. He notes that the economic potential for the region is great but so also are the competing political claims.

"THE ARCTIC OCEAN—HOT TIMES IN A COLD PLACE"

By Don Walsh, U.S. Naval Institute *Proceedings* (July 2017): 91.

Climate change is warming the Arctic twice as fast as the rest of our planet. As a result, the Arctic Ocean's highly reflective sea ice cover is decreasing quickly.

In winter, this ocean is covered completely by ice ranging from 6- to 12-feet thick. During summer months, the ice cover is reduced by half. Because of warming, however, scientists estimate that by 2040 the Arctic Ocean will be ice free in the mid-summer months. The ice loss rate has been estimated at 3 percent per decade, but this may be too conservative. In recent years the rate seems to be increasing. These numbers apply to the ocean's surface, but a more troubling number is loss of ice volume (mass). Research has found that over the past four decades the thickness of the polar ice has been reduced by 65–85 percent.

Bordered by five nations, the Arctic Ocean is the smallest, shallowest, and least known of our planet's five oceans. Despite being the smallest ocean, its surface area is one and a half times that of the continental United States. It is 17,900 feet at its deepest, although its average depth is just 3,400 feet. (World ocean average is 12,100 feet.) It is the least saline ocean because of ice melt, rivers that flow into it, and small losses of water by evaporation.

This ocean contains nearly 22 percent of the world's continental shelves. At the height of the last ice age 22,000 years ago, huge volumes of seawater were converted to ice that covered many land areas. As a result, sea level was about 460 feet lower than today. As the ice gradually melted, wave action of the rising waters pushed back shorelines forming relatively flat submarine shelves.

During continental shelf formation these shallow coastal areas received massive streams of organic and inorganic materials as runoff. This was the basis for the formation of huge seafloor hydrocarbon deposits, and added massive amounts of nutrients to support sea life. It is estimated that 25 percent of the world's undiscovered hydrocarbons are in the Arctic Ocean. It also hosts more fish species than any other ocean.

From hydrocarbons to fishery stocks and new sea routes, a seasonally ice-free Arctic Ocean could offer vast new opportunities. Russia is blessed with the world's largest continental shelf, having a maximum width of more than 1,000 miles and an area of nearly 1.5 million square miles.

The Arctic's ice cover has made it difficult to assess and exploit its resources. Explorers and scientists have been working in the Arctic since the late 19th century but knowledge of this ocean has been hard won. Today, aircraft, satellites, fixed bases on the ice, and military submarines help collect data. The research ship, however, is the most productive platform, even though working in an ice-covered sea is difficult.

In August 2007, two Russian Mir manned submersibles dove 14,000 feet to the seafloor at the geographic North Pole and planted the Russian flag.

Publicity surrounding this event, mostly in the Russian media, was somewhat breathless and implied this gave Russia vast added territorial rights. Legal experts did not take the claim seriously. But governments of the four other nations bordering the Arctic Ocean did take note, leading to decrees, promises of increased activity, and some new investments—though fewer than four million people live above the Arctic Circle.

Will the Russians eventually have rights to an extended continental shelf? Possibly. Denmark (Greenland), Norway, and Canada have made similar claims, but such claims cannot be made unilaterally. Under the U.N. Convention on the Law of the Sea treaty, an international commission will rule on seafloor claims made by all coastal nations to extend their continental shelves. While this might be politically significant, the deeper waters of the Arctic Ocean have little value in terms of resources. Accessible riches reside on the shelves.

There are really two Arctic Oceans: the political one of competing territorial claims, and the economic one with great potential resources. "Potential" is the key word. It will be some time before this wealth can be economically exploited. In the meantime, the political Arctic Ocean will continue to have vibrant activity.

18 "CHARTING THE COURSE: THE U.S. NEEDS AN ARCTIC FLEET"

CAPT Kevin Eyer, USN (Ret.)

What is the best way organizationally to effect naval operations in the Arctic? In a succinct statement of the present structure for engaging in operations, Captain Kevin Eyer calls for the reorganization of responsibilities within U.S. Fleet Forces Command and Northern Command and the establishment of an Arctic fleet similar to what occurred in 2008 with the Fourth Fleet and Southern Command.

"CHARTING THE COURSE: THE U.S. NEEDS AN ARCTIC FLEET"

By CAPT Kevin Eyer, USN (Ret.), U.S. Naval Institute *Proceedings* (April 2018): 12.

The Arctic Council's Arctic Monitoring and Assessment Program suggests the possibility that the Arctic Ocean may be virtually ice free by mid-century. In 2014, Admiral Jonathan W. Greenert, then–Chief of Naval Operations (CNO), endorsed this position when he signed the "U.S. Navy Arctic Roadmap, 2014–2030." This document outlines policy guidance, U.S. national interests in the Arctic, the regional security environment, Navy missions in the region, and the Navy's strategic objectives for the Arctic.

More important, the "Navy Ways and Means for the Near-Term, Mid-Term and Far-Term" section and Appendix 3 form what is essentially a plan of action and milestones for addressing at least the near term (2014–20) in this effort. Unfortunately, in all of this there is no organization beyond the Office of the Chief of Naval Operations, which is charged with a structural execution of the plan but with no timeframe.

The Northern Command (NorthCom) is attuned to this absence of structure and the attendant inactivity. More specifically, the Alaskan Command (AlCom), a joint subordinate unified command element of the NorthCom, is increasingly concerned with Navy inaction in the Arctic. To date, the AlCom has not been able to attain the few Navy ships necessary to demonstrate U.S. interest in the region. Certainly this is a reflection of the Navy's declining ship numbers at a time when the demand for ships is undiminished.

Normally, a combatant commander (CoCom) would route requests for forces (such as Navy ships) through its naval component commander (NCC) and then on to U.S. Fleet Forces Command (USFFC) for annual disposition. As it turns out, however, USFFC is NorthCom's NCC. It seems fair to say that this creates a difficulty for NorthCom because USFFC is senior and reports directly to the CNO. Frankly, USFFC is more interested in providing the very limited number of ships available to Navy commands such as the Fifth Fleet and Pacific Command.

This logjam must be broken. In this effort, there are two key steps. First, USFFC is too big, senior, and unresponsive for the task. This is doubly true if plans to remove major Pacific Fleet authorities and place them under USFFC go forward as is recommended. A new, separate, more junior and flexible flag—one directly responsible to NorthCom—needs to be established.

What is needed is an Arctic Fleet, and the template for this exists. In 1950, Fourth Fleet was disestablished. In 2008 the fleet was reestablished and merged with Southern Command's naval component commander to address growing interests in the Caribbean and South America. While naval forces are not permanently assigned to Fourth Fleet, the organizational structure remains in place both to support force assignment and to represent Navy interests in the region. The same can be true in the Arctic.

This new Arctic Fleet can be established in a step-wise fashion, tailored across time and married to changing force structure. A sensible first step would be to augment the small Navy staff assigned to AlCom. Subsequently, in the mid-term, a joint inter-agency task force (JIATF) could be established out of the AlCom office, as resources and activity grew. Certainly, this JIATF would include the Coast Guard, but it also should include liaison officers from Canada, Norway, and other key allies. Ultimately this fleet would be stood up and merged with the NorthCom's NCC. The Arctic Fleet could be commanded by, for example, either a Navy reserve admiral or a Coast Guard admiral.

In the interim, it seems imperative that forces be sent north to demonstrate U.S. intention and seriousness, not only to allies but also to potential adversaries. In the near term this could be a mission for LCSs, but by 2020, the start of the mid-term, a small task force should be deployed north on an annual basis to operate with Coast Guard assets and allies. That is, if the Navy is serious in its desire to gain the high ground in the far north.

19 "CHINA HAS A NEW ARCTIC POLICY"

LCDR Rachel Gosnell, USN

Lieutenant Commander Rachel Gosnell's second essay in this volume provides a succinct synopsis of China's Arctic policy and quest for a "Polar Silk Road." The economic potential in the region is enormous, and so too is the need for a U.S. Arctic strategy and policy robust enough to counter China's on every level, including enhanced naval operations.

"CHINA HAS A NEW ARCTIC POLICY"

By LCDR Rachel Gosnell, USN, U.S. Naval Institute *Proceedings Today* (May 2018).

As the ice in the High North diminishes, Arctic and non-Arctic states are seeking to explore the vast potential of the region. Congruent with its aspirations of great power status and achieving economic security, China—which calls itself a "near-Arctic" state—has furthered its Arctic interests by conducting extensive scientific research and investment in the region. Chinese interest in the Arctic has been rising for years, and the January release of a white paper entitled "China's Arctic Policy" offers insights into Beijing's Arctic ambitions.

The document articulates Chinese interests in the region and outlines its policy goals and positions on both the Arctic and participation in Arctic affairs. This white paper is a notable step, as it provides a clear guiding vision for China's involvement in the High North. Although the document repeatedly notes China's commitment to international law and respect for other nations, it asserts that China will use Arctic resources to "pursue its own interests," to include a "Polar Silk Road." This policy was explained at the recent High North Dialogue, offering additional insights into China's intent in the Arctic.

The Arctic may hold tremendous economic potential for resources and trade, with the 2008 U.S. Geologic Survey estimating the region has more than 90 billion barrels of oil, 1,700 trillion cubic feet of natural gas, and 44 billion barrels of liquid natural gas. This amounts to approximately one-third of the world's natural gas supply and 13 percent of global oil reserves. In addition, there are significant mineral resources in the region, to include bauxite, phosphate, iron, zinc, nickel, and copper. Though exploration and extraction remain constrained by the difficulty and expense of development in the region, recent years have seen China seek a greater presence in the Arctic. The Polar Research Institute of China established the Yellow River Station in the Norwegian archipelago of Svalbard in 2003 and continues to conduct research throughout the Arctic.

In 2010, Rear Admiral Yin Zhuo noted that "the Arctic belongs to all the people around the world, as no nation has sovereignty over it. . . . China must play an indispensable role in Arctic exploration as we have one-fifth of the world's population." China was granted observer status in the Arctic Council, the preeminent intergovernmental forum on the Arctic, in 2013, the same year it founded the China-Nordic Arctic Research Center. In 2014, Chinese scholar Elizabeth Economy noted that Beijing "has begun the process of engaging in the Arctic through research, investment, and diplomacy."

China's Ukrainian-built icebreaker, *Xue Long* (*Snow Dragon*) routinely performs scientific research in the Arctic. China has pursued significant investment in Greenland, Iceland, and Russia. Last June [2017], the Chinese deepwater semi-submersible rig *Nanhai-8* entered the Kola Bay, hired by Gazprom for Arctic drilling assignments. Chinese investment in energy exploration is likely to continue, particularly given the impact of Crimea sanctions on Russian relations with most Western countries. The country has focused on the region's maritime shipping routes, with President Xi unveiling the term "Polar Silk Road" last year in Moscow as part of the broader "Belt and Road Initiative." Indeed, China's second icebreaker, the *Xue Long 2* (*Snow Dragon 2*) launched from the Jiangnan Shipyard in December.

While the difficult weather and ice conditions of the region preclude large-scale maritime activity in the near future, it is clear that China is looking north for alternative trade routes. In 2013 China Ocean Shipping Company (COSCO) sent a cargo vessel along the Northern Sea Route; it sent an additional five ships in 2016 and eight the following year. Though a small fraction of overall global trade (nearly 17,600 vessels transited the Suez Canal in 2017, as compared with 24 transiting the Northern Sea Route), it indicates China's willingness to explore alternative routes. Indeed, this March saw COSCO and Russia's Novatek release a statement where officials agreed to "expand the mutual cooperation in scope and depth, especially to expand the dialogue on Arctic transportation collaboration."

The White Paper's Details

The new white paper details China's interests in the region as "an important stakeholder in Arctic affairs," and identifies specific policy goals and positions. It is divided into four main parts: "The Arctic Situation and Recent Changes," "China and the Arctic," "China's Policy Goals and Basic Principles on the Arctic," and "China's Policies and Positions on Participating in Arctic Affairs."

It first highlights the impact of global warming on the region and explores the potential impact of accelerated ice melt, addressing the region's abundant natural resources. The paper asserts that while non-Arctic states lack territorial sovereignty in the Arctic, they have rights that include scientific research, laying of submarine cables, and resource exploration and exploitation (to include mining and fishing).

Importantly, the white paper reiterates Rear Admiral Zhou's 2010 sentiment by noting that the "Arctic situation now goes beyond its original inter-Arctic States or regional nature, having a vital bearing on the interests of States outside the region and the interests of the international community as a whole, as well as on the survival, the development, and the shared future for mankind."

In opening the second part, "China and the Arctic," the paper proclaims China to be an important stakeholder in Arctic affairs. China's key areas of interest in the region are declared to be climate change, the environment, scientific research, use of shipping routes, resource exploration and exploitation, security, and global governance. This section carefully traces Chinese historical involvement in the region, beginning with the 1925 Spitsbergen Treaty and extending through its current scientific and economic pursuits in the region. It identifies the opportunities of a "Polar Silk Road" in facilitating connectivity and economic development.

The white paper next establishes Chinese policy goals as "understand, protect, develop and participate in the governance of the Arctic." It further explains each as:

- Understand the Arctic by improving capacity and capability in scientific research
- Protect the Arctic by responding to climate change and protecting the Arctic environment and indigenous peoples
- Develop the Arctic by improving Chinese capacity and capability, to include the development of Arctic shipping routes
- Participate in the governance of the Arctic by commitment to the existing framework of international law, including the U.N. Charter, the U.N. Convention on the Law of the Sea (UNCLOS), the International Maritime Organization (IMO), and climate change and environment treaties

While the paper notes that these goals will be accomplished by working with other countries, it affirms that China will pursue its own interests and address security threats through global, regional, multilateral, and bilateral mechanisms.

To achieve the stated policy goals, the paper declares that China will participate in Arctic affairs according to the principles of: "respect, cooperation, win-win result and sustainability."

- *Respect* is defined as abiding by international treaties and respecting the rights of Arctic and non-Arctic states and the interests of the international community.
- *Cooperation* applies to all stakeholders, intergovernmental organizations, and non-state entities, with the paper noting that cooperation is necessary on the environment, scientific research, shipping route development, resource utilization, and cultural activities.
- *Win-win result* is noted as the pursuit of mutual benefit and common progress.
- *Sustainability* is highlighted as the fundamental goal of China's Arctic involvement, such that promoting sustainable development in the Arctic will be critical.

Why the International Community Should Care

While the white paper addresses China's desire to cooperate and act in accordance with established international law and international treaties, it is important that strategic interests are woven subtly throughout the document. The paper clearly expresses Chinese intent to pursue a Polar Silk Road for maritime shipping traffic. Indeed, this harkens to the 2016 observation by the COSCO Executive Vice President Ding Nong, who noted that "as the climate becomes warmer and polar ice melts faster, the Northeast Passage has appeared as a new trunk route connecting Asia and Europe." The white paper emphasizes the Polar Silk Road and the need for new icebreakers.

The United States and the rest of the international community should welcome China's stated commitment to "respect, cooperation, win-win result and sustainability." Indeed, the white paper uses the words *cooperation* 45 times and *respect* 23 times, though it remains to be seen if China will uphold this commitment. Indeed, at the 2018 High North Dialogue panel on "China's Future Role in the Arctic," Liu Jun, professor and dean at the School of Advanced International and Area Studies of East China Normal University, displayed a slide that highlighted the number of times specific words were included in the document—to include *cooperation* and *respect*. Clearly the Chinese wanted the audience to be aware of this point.

The white paper cements China's belief in its "stakeholder" role in the Arctic; *interests* are cited 18 times although *security* is used only 10 times. China has invested significantly in the region, and the white paper indicates the trend will continue. Crimea-related economic sanctions likely have increased the cooperation between China and Russia in the Arctic. Notably, the state-owned PetroChina has a 20 percent stake in Russia's Yamal LNG project; indeed, the first LNG shipment from Yamal in December went to China. The white paper makes it clear that these activities and investments will continue as China seeks a greater role in the Arctic.

China's white paper also points out that current U.S. Arctic policy lacks robustness and must be strengthened to ensure future U.S. strategic interests are achievable. The white paper mentions the UNCLOS ten times—a treaty that the United States has not ratified but one that China and 167 other states have. China has expressed throughout the white paper that it will abide by UNCLOS. Without ratifying UNCLOS, the United States has no legal authority with which to pursue claims and ensure consistent application of international law, potentially diminishing U.S. standing and disadvantaging economic interests.

Overall, the codified Chinese Arctic strategy serves to highlight the shortfalls in U.S. Arctic policy and the need for U.S. policymakers to reexamine the Arctic to ensure protection of strategic interests. The issuance of this white paper by Beijing clearly signifies China's intent to play a significant role in the Arctic's future—the United States should be prepared to do the same.

20 "WORLD NAVAL DEVELOPMENTS: ARCTIC ACTIVITY HEATS UP"

Dr. Norman Friedman

Norman Friedman, a prominent naval analyst, provides an overview of international naval activities in the Arctic and reminds readers that, although Chinese naval operations in the South China Sea and Russian operations in the Black Sea and the Sea of Azov may capture headlines, the United States must not overlook the Arctic. Both China and Russia have expressed growing interest in the region. Both nations remain a focus of U.S. strategic attention, and both nations may soon act in a manner that requires more consistent and extensive operations by the U.S. Navy.

"WORLD NAVAL DEVELOPMENTS: ARCTIC ACTIVITY HEATS UP"

By Dr. Norman Friedman, U.S. Naval Institute *Proceedings* (November 2013): 90–91.

From a naval perspective, perhaps the single most significant impact of global warming is the transformation of the Arctic from a forbidding, ice-blocked region usable almost exclusively by nuclear submarines, to a future set of sea routes and an important area of natural resources. The latter applies not only to the sea floor but also to the bordering land, which is nearly empty because the climate has been so brutal for much of the year. Now retreating ice and a warmer Arctic climate make it possible to envision the regular use of the passages around Canada and the Arctic coast of Alaska, and similarly around the northern coast of Russia—for North America, the fabled Northwest Passage sought unsuccessfully by so many explorers in the past.

Warming may also make the lands bordering the Arctic Ocean much more usable. For example, more of Canada's population may choose to live in the north. For some years the Canadians have been talking about their "third coast" and have recently revived northern land patrols as a way of asserting sovereignty over their northern territories. A warming and more usable Arctic may also improve the status and economics of indigenous peoples such as the Alaskan Aleuts. The U.S. Coast Guard has been discussing the implications of global warming and a more open Arctic for Alaska.

The Arctic Ocean itself is relatively shallow (average depth is 3,281 feet, which means that many areas are much shallower; the average depth of the World Ocean is 12,454 feet). That is why in the past even submarine navigation carried considerable hazards, as a submarine could be caught between ice and the bottom. If it becomes more hospitable, shallow depth may make seabed mining or drilling for oil relatively easy. Most of the other accessible shallow sea areas are already being exploited. The Arctic seabed is untapped, except at its edges. No one can be sure of what resources, if any, lie under the Arctic Ocean, but there seems to be a general assumption that they are likely to be vast and valuable.

Russia Returns

In September [2013], the Russians sought to cement their claims to the resources of the Arctic Ocean by steaming a group of warships, including their remaining nuclear-powered cruiser *Petr Velikiy* and two large amphibious ships, 1,500 nautical miles through the Barents and Kara Seas. The Russians described the operation as a demonstration of their legal right to the resources of the Arctic, a process that seems to have begun a few years ago when they planted a metal flag on the floor of the Arctic Ocean. The ships carried material to be used in reopening a Cold War–era Arctic air base on Kotelny Island in the Novaya Zemlya Archipelago. This archipelago was used regularly for nuclear tests, which the airfield presumably supported.

The Russian surface group was supported by all three of their operational nuclear-powered icebreakers. At present the Russians have by far the world's largest fleet of icebreakers, including six with nuclear power, and they are building more. In addition, there are nine large conventionally powered ocean icebreakers, and many smaller ones. The U.S. Coast Guard has three icebreakers, two of which are more than 30 years old, and procurement of replacements has been deferred due to projected cost. Canada has two large and four medium icebreakers, plus light icebreakers not suited to Arctic waters.

Many of the large Russian icebreakers were built during the Cold War, when the most important Arctic resource, in the former Soviet Union's view, was the potential for military basing. Some of the nuclear icebreakers used the same reactors that were mass-produced for submarines. The Soviets' Northern Fleet, based in the Western Arctic, was by far their most important. Some Soviet ballistic-missile submarines were conceived specifically to operate under Arctic ice, firing their missiles through the naturally occurring openings (polynyas). This practice was why the U.S. Navy and the Royal Navy regularly practiced under-ice submarine operations.

The Russian announcement of the Arctic operation claimed that it was unprecedented, that at no time in the modern history of Russia had surface ships penetrated so far along the Northern Sea Route (the route between the Barents and the Pacific). However, the Russians officially opened the route in 1933, having set up a special administrative agency to oversee it, and in 1935 warships of the Baltic Fleet made the passage to strengthen Soviet Pacific forces during a crisis with Japan. In July–September 1940 the German commerce raider *Komet* passed from the Barents to the Pacific via the Bering Strait between the Soviet Union and Alaska. Soviet help was essential. The voyage was considered a remarkable feat at the time. Later the Arctic Sea Route, particularly its eastern end, was an important means of shipping prisoners into some of the bleakest parts of the Soviet Gulag system.

The Russians announced the September operation as a step toward maintaining their status as (in their view) the leading Arctic power, with a combination of vital security and economic interests in the region. They pointed to claims by other, non-Arctic, countries such as China, India, and Brazil (but did not mention more realistic claims by Canada and the United States). Of these, China most recently demonstrated interest in the Arctic in 2012 when its icebreaker *Xue Long* crossed the Arctic Ocean from the Bering Strait to Iceland and back. A second Chinese icebreaker is being built. Brazil has one polar scientific ship, but she is not equipped to break anything beyond thin ice. India is in the process of acquiring an icebreaker for polar research, and already has a floating ice research station in the Arctic.

China's Interest

It seems odd that the Russians appear to have forgotten their previous use of the Northern Sea Route. If they do not believe they can operate freely along its full length, they may find it difficult to protect their sovereignty in the area. Their mention of China's interest in the Arctic may imply a fear that at some point the Chinese will see the eastern end of the old Northern Sea Route as a means of outflanking Russian forces in eastern Siberia. The Russians are well aware that China needs more and more resources as it modernizes. Although some of those resources come from Russia (i.e., Siberia), the bulk come by sea. Indeed, some Chinese naval officers have argued publicly that China now depends so much on seaborne resources that it must have a navy capable of protecting that vital interest. It is not clear to what extent this argument has caused the Chinese government to invest in a fleet; the current upsurge in naval spending may merely be part of a larger program of military modernization.

If, however, the current programs do reflect an economic argument, then the ground forces are probably also looking for justification. They could argue that China must be prepared to defend its stake in Siberian resources. This could be justified in terms of the longstanding Chinese claim that tsarist Russia seized Siberia from Imperial China through a series of "unequal treaties" extending over several centuries. At one time many Chinese lived in what is now Siberia; the Russians ejected all of them. In the 1980s the Chinese wanted the Soviets to acknowledge their territorial sins, but they admitted that since no ethnic Chinese remained in Siberia, it was difficult to justify territorial demands. Now, however, large numbers of ethnic Chinese live there. If the climate keeps improving, more will probably move in, attracted by the open spaces available for farming.

Russia is gradually modernizing its forces, but it is doubtful that there is much money for new infrastructure. What existed at the end of the Cold War is what survives today; the revived airfield in the Arctic is a good example. Through the end of the Cold War, the Soviets invested heavily in the defense of Siberia against a possible Chinese attack, expected to come mainly overland, from the south. At the time the Chinese had little naval capability. The naval side of Siberian defense against China was mainly a matter of using naval forces to outflank Chinese ground units headed north. The Soviets invested heavily in air-cushion landing craft, which were often associated with the Far East. It is unlikely that much effort went into the defense of the Arctic coast of Siberia. Little was probably done in the way of road or rail connections from garrisons nearer the Pacific coast to areas farther north and west.

Right now, the Arctic is still inhospitable enough that no one in Beijing is likely to contemplate moving troops up through the Bering Strait and onto the Arctic shore of Siberia. However, the situation is visibly changing. If that continues, the Russian Arctic shore will become a potential area of military operations, accessible to the amphibious ships the Chinese are building, which up to now have been associated with a possible assault on Taiwan. Even if the Chinese never mount an operation against Siberia, the sense that they have a growing capability to do so would, it seems, insure against any attempt by the Russians to use raw material supply to apply pressure as they have done against Europe using natural gas supplies.

At present Sino-Russian relations are said to be good. The Russians are continuing to sell China modern military hardware, although that trade is falling off as the Chinese develop their own alternatives. China has joined Russia in the Shanghai Cooperation Organization (the other members are former

Soviet republics Kazakhstan, Kyrgystan, Tajikistan, and Uzhbekistan). In theory the organization coordinates both economic and military relations. Meetings of the Shanghai organization are sometimes used to announce Russian military initiatives, such as the decision to revive long-range bomber patrols toward NATO areas. The Chinese delegates have so far refrained from any impolite reminders of past Russian sins—but those may yet come.

21 "RUSSIA OPENS ITS MARITIME ARCTIC"

CAPT Lawson W. Brigham, USCG (Ret.)

Russian activity and interest in the Arctic is growing daily. Russia's rich natural resources and its desire to increase the productivity, commercial activity, and transportation capability of its Arctic coastal region have caused the nation to pursue an aggressive program of strategic development that has gained worldwide attention since 2010. Captain Lawson W. Brigham's article provides a valuable overview of Russian activity as it was almost a decade ago. The article makes a strong argument that Russia is committed to being the preeminent and most prominent power in the region and that it will seek to direct and dominate Arctic affairs through alliances, infrastructure, and other means, even though Russia is integrated with the global community.

"RUSSIA OPENS ITS MARITIME ARCTIC"

By CAPT Lawson W. Brigham, USCG (Ret.), U.S. Naval Institute *Proceedings* (May 2011): 50–54.

As use of the Russian Arctic coastal seas expands and commercial interests drive marine transportation along the Northern Sea Route, the region is linked increasingly to the rest of the planet. Natural-resource developments in these northern onshore and offshore areas are closely tied to the future of the Russian Federation, as higher global commodity prices spur exploration and new investments in Russia's Arctic infrastructure.

The nation has developed a program for strategic development of the region, in recent pronouncements promoting Arctic cooperation as a central theme. Diplomatic developments and marine operations during 2010 have also aroused worldwide attention to this formerly remote and closed region of the Soviet era.

Barents Sea Agreement

After 40 years of negotiating, Norway and Russia announced in April 2010 that a preliminary agreement had been reached on maritime delimitation and cooperation in the Barents Sea and Arctic Ocean.[1] The differences in boundary lines between the two Arctic states in the Barents Sea (and by extension north into the Arctic Ocean) had remained problematic, but broad Norwegian-Russian fisheries cooperation in the region has existed since 1975. Recent pressures for expanded oil and gas exploration in and near the disputed areas made the lack of a boundary agreement more vexing.

The new treaty concerning "Maritime Delimitation and Cooperation in the Barents Sea and the Arctic Ocean" was signed 15 September 2010 in Murmansk by Russian and Norwegian foreign ministers Sergey Lavrov and Jonas Gahr Støre. It is historic in several ways. Not only does it establish a stable and secure Arctic boundary, it also includes detailed annexes addressing fisheries and trans-boundary hydrocarbon deposits. Both nations noted the importance of close Arctic fisheries cooperation and agreed that the Norwegian-Russian Joint Fisheries Commission will continue to handle the negotiation of total allowable catches and quotas, while considering measures such as monitoring and control related to jointly managed fish stocks.

Annex II addresses the complicated issue of a hydrocarbon deposit extending across the new boundary. A joint operating agreement will now be required to explore and exploit, as a single unit, any trans-boundary deposit. Norway and Russia also agreed to establish a joint commission for consultations, exchange of information, and as a means of resolving issues.

The culmination of this significant accord, once it has been ratified by the two parliaments, will strengthen Norwegian-Russian cooperation in a key Arctic maritime region and remove a longstanding, disputed area from Arctic state concern. For the Russian Federation and Norway, this agreement provides a framework of cooperation and a stable political environment in which the Barents Sea's continental-shelf hydrocarbon resources can be increasingly exploited. The treaty also provides a unique and workable model for further circumpolar cooperation.

Trans-Arctic Voyages and Shuttle Operations

The Northern Sea Route, defined in Russian federal law as the set of waterways from Kara Gate (southern tip of Novaya Zemlya) to the Bering Strait, does not include the Barents Sea. The navigation season of 2010 for this route was notable not for total tonnage carried or number of ships, but for several experimental trans-Arctic voyages involving diverse ship types. Four of the voyages took place during the summer, when sea ice is at its minimum in August and September; the fifth was a historic east-to-west escort of an icebreaking offshore vessel in December.

Sovcomflot's ice-class tanker SCF *Balnca* (Liberian flag) completed a voyage carrying gas condensate from Murmansk to Ningbo, China, in 22 days; a reduced draft and slower speeds were necessary through the shallow straits of the New Siberian islands.[2] SCF *Baltica* is the first tanker of more than 100,000 deadweight tons to sail the Northern Sea Route, testing its viability for high tonnage. Also testing the route was the *Nordic Barents* (Hong Kong flag), an ice-class bulk carrier, on a voyage with iron ore from Kirkenes, Norway, to China. This was the first foreign-flag ship to carry cargo from one non-Russian port to another through Russian Arctic waters.[3] The route has the potential to link northern European mines to markets in China, Japan, Korea, and other Pacific nations.

In a similar voyage, Norilsk Nickel's icebreaking carrier *Monchegorsk* sailed from Murmansk and Dudinka along the Northern Sea Route east to Shanghai.[4] However, the key difference in comparison with other full transits was that this one was conducted by an ice-capable commercial ship sailing the length of the route without icebreaker escort. With a change in federal regulations, such independent sailings could become more common during the short summer navigation season.

Two 2010 voyages were unique. On 28 August the passenger ferry *Georg Ots* departed St. Petersburg for Murmansk and a subsequent voyage under nuclear icebreaker escort along the Northern Sea Route, arriving in Anadyr, Chukota, on 26 September. The ferry reached its new home port of Vladivostok in October, for use during the 2012 Asian-Pacific Cooperation Summit and future local operations.[5] More challenging was the 16–26 December escort by the nuclear icebreaker *Rossiya* of the icebreaking offshore vessel *Tor Viking* from the Bering Strait to the northern tip of Novaya Zemlya across the Northern Sea Route.[6] This successful voyage indicates the sailing season may be extended for passage of ice-capable ships under close escort.

Arctic shuttle operations are the key to efficient marine transportation of natural resources in the Barents and Kara seas, encompassing the western end of the Russian maritime Arctic. Two innovative systems are fully developed and operate year-round. A five-ship Arctic icebreaking carrier fleet carries nickel plate from Dudinka on the Yenisey River to Murmansk; this fleet is owned and operated by Norilsk Nickel, the mining complex in western Siberia, and year-round navigation has been maintained since 1979.

A three-ship icebreaking tanker operation services the offshore oil terminal at Varandey in the Pechora Sea (southeast corner of the Barents Sea). The three Panamax-size shuttle tankers can annually deliver nearly 12 million tons of oil to a floating tank farm in Murmansk.[7] The terminal and marine shuttle system represent a prime example of Arctic globalization: the Russian company Lukoil teamed with the American firm ConocoPhillips for investment and development of the offshore terminal; the tankers were built in Korea by Samsung Heavy Industries using Finnish icebreaking technology; and the ships are operated by Sovcomflot.

A third shuttle system is due for full operation later in 2011; a two-ship fleet will deliver oil to Murmansk from the Prirazlomnoye offshore oil production platform in the Pechora Sea.[8] Both tanker shuttle fleets have significant potential to provide year-round service to other projects and thereby optimize regional marine operations.

China and Finland Alliances

As hydrocarbon exploration and transportation development of the Russian maritime Arctic have rapidly evolved, Russia has been quick to forge strategic commercial alliances with China, as well as Finland and other Western companies. Early in the operation of the Varandey terminal, Lukoil signed an agreement with Sinopec (China Petroleum and Chemical Corporation) to supply 3 million tons of oil to China.[9]

Sovcomflot Group reported on 22 November 2010 that it had signed a long-term agreement with China National Petroleum Corporation regarding seaborne carriage of hydrocarbons from the Arctic to China. The cooperative agreement envisions using the Northern Sea Route for not only moving oil and gas from Russia's developing offshore, but also for trans-Arctic shipments in the summer navigation season. It includes a provision for Sovcomflot to assist in the training of Chinese mariners in Arctic navigation.[10]

A new venture was created between Russian and Finnish commercial interests in December 2010. STX Finland Oy and the United Shipbuilding Corporation (composed of 42 shipyards in Russia) formed a joint venture that will focus on Arctic shipbuilding technology. The newly named Arctech Helsinki Shipyard Oy will build specialized icebreaking vessels for key operators throughout the Russian maritime Arctic, and likely also for foreign buyers.[11]

Arctic Hub and Infrastructure

The ice-free port of Murmansk has long been viewed as a critical economic component of the Russian maritime Arctic. Recent reports in Russia confirm a strategy to fully develop Murmansk as the major oil, gas, and container port, as well as a transportation hub for the entire Russian Arctic. Tax and customs benefits from a new port economic zone will facilitate investment, as Murmansk is increasingly tied to offshore development in the Barents Sea.[12] Companies such as BP, for its potential Kara Sea venture; and others such as Gazprom, planning the offshore Shtokman gas field, look to establish bases for Arctic operations (including response and emergency services) in Murmansk.

Northern Sea Route headquarters of the western sector may be moved from Dikson, on the remote Kara Sea coast, to Murmansk. As well, it is clear that new port and construction activities along the Russian Arctic will be serviced from a modern hub in Murmansk. More new marine infrastructure has been planned. New Arctic rescue centers, Russian-built satellite systems for the North, and a new Arctic research vessel were all discussed in 2010 by several federal ministries. Some of this critical Arctic infrastructure may come about through investment by public-private partnerships, including foreign capital.

The Russian nuclear-powered icebreaker fleet under the state-owned Atomflot (part of Rosatom) is a legacy of the Soviet Union, but retains near iconic status in the Russian north and the polar world. There are plans to modernize the fleet by building dual-draft ships that can operate along the coastal waters of the Northern Sea Route and in the Siberian estuaries and rivers. It is apparent that shuttle fleets in the Barents and Kara seas do not intend to operate with icebreaker support or in convoys. However, the nuclear icebreakers would be used to escort Russian and foreign ships along the Northern Sea Route during extended navigation seasons and to conduct scientific expeditions, support Arctic oil and gas offshore development, and support summer sealift to Arctic communities. Most certainly the nuclear icebreakers remain a visible and tangible presence of the Russian Federation in the Arctic Ocean.

State Policy and International Cooperation

Russian President Dmitry Medvedev approved a new Arctic policy statement on 18 September 2008, titled *The Foundations of the Russian Federation's State Policy in the Arctic until 2020*. This document outlines the strategic priorities for the Russian Federation in the Arctic, noting unique features of the region including low population, remoteness from major industrial centers, a large natural-resource base, and dependence on supplies from other regions in Russia. One of the critical points is that Russia intends to use its Arctic regions as a "strategic resource base."

For the maritime world, the policy mentions use of the Northern Sea Route as a national, integrated "transport-communications system" in the Arctic, specifically an "active coast guard system" in the Russian Arctic under the direction of the Federal Security Service. Important for the Arctic states, the document notes Russia's interest in enhancing cooperation with other national coast guards in the areas of terrorism on the high seas, prevention of illegal immigration and smuggling, and protection of marine living resources. Russia, Norway, and the United States already cooperate in these pursuits, but more can be expected as marine activities expand throughout the Arctic Ocean.

On 22–23 September 2010, the Russian Geographical Society held a key conference in Moscow that focused on the importance of international cooperation. Appropriately called "The Arctic: Territory of Dialogue," this forum gave prominence to the roles of indigenous people, the need to protect the environment, the vast storehouse of Arctic resources to be developed, and the need to affirm the region as a "zone of peace and cooperation." Prime Minister Vladimir Putin addressed the conference in a wide-ranging speech, noting that 70 percent of the country is located in northern latitudes, and that the issues of Arctic development are high on Russia's national agenda. He mentioned the importance of the Arctic Council to the "integration" of ideas and concepts, as well as the upcoming search and rescue agreement to be signed by the Arctic ministers in Nuuk, Greenland, this May [2011].[13]

Overall Implications

The Russian Federation is embarking on a long-term strategy to link its Arctic region economically to the rest of the globe. The drivers are clearly the development of natural resources and timely export of domestic production. The facilitators are innovative marine transportation systems that can move cargoes of hydrocarbons and hard minerals both westbound (year-round) and eastbound (summer season) along the top of Eurasia.

There will be opportunities for ice-capable, foreign-flag ships to gain access to Russian Arctic waters, as illustrated by recent operations in summers 2009 and 2010. For example, bulk carriers could increasingly link northern European mines to Pacific ports during summer seasons of navigation. And foreign-flag ice-class tankers could compete with modern Russian-owned fleets of advanced carriers for this potential summer maritime trade route, especially linking China to Russian Arctic oil and gas.

For safety and security reasons, Russia is sure to manage tightly the opening of its Arctic waters to maritime trade. Similarly, the capabilities of its border guard of the Federal Security Service will be enhanced for Arctic operations. There has been no change in regulations along the Northern Sea Route for mandatory icebreaker escort in certain straits, despite Norilsk Nickel's *Monchegorsk* full passage without escort in 2010. Non-ice-strengthened commercial ships have not yet sailed along the eastern reaches of the route.

Changes could come soon, with legislative action from the State Duma. All this new activity will require improved environmental observations, new marine charts, traffic monitoring, enforcement capability, and control measures. We are witnessing the cautious evolution of an Arctic region from a once-closed security bastion to a vast marine area more open for use and, potentially, integrated with the global economy.

Notes

1. Press release, 15 September 2010, Office of the Prime Minister, Norway, http://www.regjeringen.no/en/dep/smk/press-center/Press-releases/2010/treaty.html?id=614254.

2. Press release, 9 September 2010, SCF (Sovcomflot) Group, Russia's largest shipping company (state owned), http://www.scf-group.com.

3. Joint press release, 20 August 2010, Nordic Bulk Carrier A/S and Tschudi Shipping Company A/S.

4. Press report, 16 November 2010, Norilsk Nickel, on the return of its carrier *Monchegorsk* to the port of Dudinka on the Yenisey River from Shanghai, http://www.nornik.ru/en/press/news/3101/.

5. Reported in *MB News* (Murmansk Business News), 29 September 2010; and *BarentsObserver.com*, 30 September 2010.

6. Voyage details provided by the Swedish shipping firm TransAtlantic, operator of *Tor Viking*, January 2011.

7. *Oil and Gas Eurasia*, "First Shipment of Oil from the Varandey Terminal," 10 June 2008, http://www.oilandgaseurasia.com/news/p/0/news/2444/.

8. *Oil and Gas Eurasia*, "Double Acting Tankers for the Prirazlomnoye Project," November 2010, http://www.oilandgaseurasia.com/articles/p/130/article /1373/.

9. Reuters, Moscow, 16 June 2009, from a Lukoil report.

10. Press release, 22 November 2010, Sovcomflot Group.

11. Press release, 10 December 2010, STX Finland and United Shipbuilding Corporation, agreement signed in St. Petersburg.

12. "Planned Seaport Would Turn Murmansk into Major Hub," *St. Petersburg [Russia] Times*, 5 October 2010.

13. Address to the international forum, "*The Arctic: Territory of Dialogue*," 22 September 2010, official site of the prime minister of the Russian Federation, http://premier.gov.ru/eng/events/news/12304/.

22 "AMERICA'S ARCTIC IMPERATIVE"

ADM Robert J. Papp Jr., USCG (Ret.)

This anthology began with an article by Admiral Robert Papp Jr., and appropriately it concludes with one. This short essay is an exhortation for the United States to continue its presence in the Arctic and its global leadership of nations, as changes in the environment and economic interests continue to develop rapidly. In the midst of these changes, maritime and naval operations will accelerate, and the U.S. Navy and U.S. Coast Guard must be prepared to meet the operational challenges.

"AMERICA'S ARCTIC IMPERATIVE"

By ADM Robert J. Papp Jr., USCG (Ret.), U.S. Naval Institute
Proceedings (February 2017): 48–49.

During my nearly 40 years in the U.S. Coast Guard, the Arctic was a common thread. My first tour serving in a cutter homeported at Adak, Alaska, gave me an intimate knowledge of navigating dangerous Arctic waters, of the isolation of living on a distant Aleutian Island, the cooperative nature of our relationship with Russia along Alaska's western maritime boundary, and the difficulties of carrying out Arctic search-and-rescue missions. But while I came to understand the importance of the Arctic through first-hand experiences, I asked myself, what will compel our citizens to care about the region?

During my two years as U.S. Special Representative to the Arctic, I had the opportunity to meet and speak with people living and laboring in the region and those working hard on its behalf. Whether a lifelong Arctic resident, a national Arctic policy maker, a knowledgeable academician, or one of our U.S. Arctic Youth Ambassadors, each genuinely conveyed to me the importance of the Arctic region to our nation. Time and again, they expressed concerns about the impact of climate change upon it.

Across Alaska, the impacts of climate change are plainly evident and undeniable: coastal erosion is causing homes to fall into the sea or other waterways; thawing permafrost is affecting infrastructure; and unstable ice and changing migration patterns are preventing people from obtaining the food sources they have relied upon for tens of thousands of years.

Hearing Gwich'in community members share how climate change has impacted the safety and security of Fort Yukon's villagers brings home the real impact of our changing environment. They tell of how the once-trusted trails that cross the frozen ice of the Yukon River, which they often take to reach friends, family, and hunting and gathering grounds, are now unpredictable, and more people are falling through the ice and drowning. In Barrow, the underground utility tunnel that supplies electricity to the entire city is threatened by coastal erosion. If the tunnel floods—and it will—the community will be without critical infrastructure that is taken for granted by most in the United States.

As Secretary of State John F. Kerry said when he visited Alaska in 2015, "climate is not a distant threat for our children and their children to worry about. It is now. It is happening now. The Arctic has never been an easy place to survive but global climate change now threatens life in this region in a way that it hasn't been threatened for 10,000 years."

Over the past two years [2015–17], the United States has shown great leadership in the Arctic. We have initiated vigorous local-, state-, and national-level conversations about the unique challenges and opportunities in the Arctic. This enhanced dialogue is revealing innovative paths forward and identifying areas that still require greater attention and understanding. The U.S. Chairmanship of the Arctic Council has proven a strong catalyst and focused the attention of our federal agencies, embassies, the private sector, and others on America's Arctic like never before. The Arctic Council itself—which at its 20-year mark is a brighter beacon than ever for international cooperation and policy making in the region—has been strengthened through our chairmanship. President Barack Obama established the Arctic Executive Steering Committee to ensure that our federal efforts in the region are more efficient, effective, and coordinated. The Arctic Coast Guard Forum and operational field exercises have better prepared our first responders for maritime search-and-rescue and oil pollution response.

The willingness of stakeholders to engage in a constructive dialogue has helped us to create unparalleled momentum around, action on, and commitment to Arctic issues in this country. Every meeting, public engagement, and media program were opportunities to learn from, draw attention to, and gain consensus on how to move forward. From our most recent Arctic Council gathering in Portland, Maine—one of the first official Council meetings held outside the Arctic—to the Conference on Global Leadership in the Arctic: Cooperation, Innovation, Engagement, and Resilience, which brought a sitting U.S. President to the U.S. Arctic for the first time, our country is experiencing an unprecedented level of engagement and exchange on Arctic issues.

In his January 1961 farewell speech to the Commonwealth of Massachusetts, not long before he assumed the presidency, John F. Kennedy reminded us of Pericles' words to the Athenians, "We do not imitate—for we are a model to others." Kennedy continued, stating, "Today the eyes of all people are truly upon us."

The United States must continue demonstrating great leadership in the Arctic because the world's eyes are upon us as never before. The urgency of climate change and the resultant increase in human activity make our intentions and actions in the region matter more than ever. Now is our opportunity to show the world that we as a nation have the tenacity, capabilities, and determination to meet the economic, security, and environmental needs of the region and its peoples. The work we have started together on the Arctic and the domestic and international commitments we have made to this vital region must continue.

INDEX

ABOUT THE EDITOR

Timothy J. Demy is a professor of military ethics at the U.S. Naval War College and served in the Navy for twenty-seven years. He has been to the Arctic several times, has done extensive research at the Scott Polar Research Institute in Cambridge, UK, and is a member of the Arctic Studies Group at the Naval War College.

The Naval Institute Press is the book-publishing arm of the U.S. Naval Institute, a private, nonprofit, membership society for sea service professionals and others who share an interest in naval and maritime affairs. Established in 1873 at the U.S. Naval Academy in Annapolis, Maryland, where its offices remain today, the Naval Institute has members worldwide.

Members of the Naval Institute support the education programs of the society and receive the influential monthly magazine *Proceedings* or the colorful bimonthly magazine *Naval History* and discounts on fine nautical prints and on ship and aircraft photos. They also have access to the transcripts of the Institute's Oral History Program and get discounted admission to any of the Institute-sponsored seminars offered around the country.

The Naval Institute's book-publishing program, begun in 1898 with basic guides to naval practices, has broadened its scope to include books of more general interest. Now the Naval Institute Press publishes about seventy titles each year, ranging from how-to books on boating and navigation to battle histories, biographies, ship and aircraft guides, and novels. Institute members receive significant discounts on the Press's more than eight hundred books in print.

Full-time students are eligible for special half-price membership rates. Life memberships are also available.

For a free catalog describing Naval Institute Press books currently available, and for further information about joining the U.S. Naval Institute, please write to:

Member Services
U.S. NAVAL INSTITUTE
291 Wood Road
Annapolis, MD 21402-5034
Telephone: (800) 233-8764
Fax: (410) 571-1703
Web address: www.usni.org